D0228976

30131 05750758 1

LONDON BOROUGH OF BARNET

EX LIBRIS

TEST GODS

Tragedy and Triumph in the
New Space Race

NICHOLAS SCHMIDLE

HUTCHINSON
LONDON

1 3 5 7 9 10 8 6 4 2

Hutchinson
20 Vauxhall Bridge Road
London SW1V 2SA

Hutchinson is part of the Penguin Random House group of companies whose
addresses can be found at global.penguinrandomhouse.com.

Penguin
Random House
UK

First published in the United States by Henry Holt in 2021
First published in the United Kingdom by Hutchinson in 2021

www.penguin.co.uk

A CIP catalogue record for this book is available from the British Library.

ISBN 9781786331854 (hardback)
ISBN 9781786331861 (trade paperback)

Printed and bound in Great Britain by Clays Ltd, Elcograf S.p.A.

The authorised representative in the EEA is Penguin Random House Ireland,
Morrison Chambers, 32 Nassau Street, Dublin D02 YH68.

Penguin Random House is committed to a sustainable future for
our business, our readers and our planet. This book is made from
Forest Stewardship Council ® certified paper.

To Oscar and Bohan

"*They were either the end of the old or the first of the new.*"

–NORMAN MAILER, *OF A FIRE ON THE MOON*

"*All men live enveloped in whale lines.*"

–HERMAN MELVILLE, *MOBY-DICK*

CONTENTS

TEST GODS

PROLOGUE

BLUE ZEBRA

A SPLIT SECOND INTO THE MISSION Mark Stucky knew something was horribly wrong. Pushing the stick forward, he had expected to enter an aggressive dive, like a kamikaze bomber racing at its target—in this case the bleak California desert. But now the tail of his spaceship was stalled and beginning to drift, contorting his carefully calibrated dive into an unintended back flop.

The computer on board the spaceship was going berserk—alerts beeping, yellow and red lights flashing. Grunting, Stucky pulled on the stick to try to level out. Nothing happened. He was now upside down and floating out of his seat, 40,000 feet in the air. The straps of his harness dug into his shoulders. The ship was falling fast.

Think.

An average human brain weighs about three pounds and contains nearly a hundred billion neurons; an almond-shaped cluster near the brain stem handles our response to fear. Most people panic when they're afraid. Their palms sweat, their hearts pound, and their minds freeze—at the exact moment acuity is needed most.

Stucky was not most people.

He thumbed the pitch trim switch, hoping the pair of horizontal stabilizers on the tail booms would bite the air. No response. He reached up and switched to the emergency trim system. No response.

Already upside down, now the spaceship was beginning to spin. Stucky counted each rotation as the plunging craft spun past the sun.

One . . . two . . .

Stucky remained almost mysteriously calm. Clinical. He found an odd sort of comfort in such moments. His job was dangerous enough without letting panic get in the way. He was a test pilot, determined to navigate unexplored aerodynamic realms so that his engineering colleagues could define the spaceship's capabilities and limits; as Arthur C. Clarke said, "The only way of discovering the limits of the possible is to venture a little way past them into the impossible."

Each test flight offered some new adventure. But "expanding the envelope," as test pilots described their work, was not adventurism for its own sake. It was a methodical process that drew as much on the discipline and rigor of the scientist as on the artful improvisation of the daredevil.

Fly, test, notate, adjust; fly, test, notate, adjust.

Stucky rummaged through a mental catalog of personal experiences and training manuals and anything he'd ever read or heard from any other pilot in search of something useful, some way to save his ship—and his life.

He deployed the speed brakes. Nothing. Stepped on the opposite rudder pedal. Nothing. The spaceship continued to tumble and corkscrew at an alarming rate, losing 1,000 feet of altitude every two seconds. The sun kept flashing in the cockpit windows.

Three . . . four . . .

"We're in a left spin," his copilot, Clint Nichols, announced over the radio, his voice flat as a clerk requesting a cleanup on aisle four.

Stucky had practiced entering and recovering from inverted spins like this plenty of times in other crafts. They were nonetheless unpleasant and dangerous maneuvers. In 1953, Chuck Yeager was flying an X-1—the same type of rocket ship he used to break the sound barrier—when he entered an inverted spin at 80,000 feet and spent nearly a minute "fighting to try to recover the airplane and stay conscious from the high rotational rates." He eventually regained control, at 25,000 feet. Thirty-two years later, the stunt pilot who filmed Yeager's scenes in *The Right Stuff* was doing stunts for the movie *Top Gun* when he got into an inverted spin, crashed, and died.

Stucky was confused: he couldn't understand why the tail had stalled. Stumped, he felt sickened that the last option to avoid an almost certain death was going to require him to unbuckle, crawl down, open the hatch, jump out, throw his parachute, and watch as Richard Branson's multimillion-dollar spaceship smashed into pieces on the desert floor, and, perhaps with it, Branson's dream of making his space tourism company, Virgin Galactic, a reality.

Stucky was chasing his own dream. He'd spent almost forty years trying to become an astronaut. He'd done stints in the Marines, the Air Force, and NASA, and he now worked for an experimental aviation firm, Scaled Composites, which Branson, a showboating British mogul, had hired to build and test a spaceship for commercial use. It was beyond zany, Branson's dream of sending passengers into space aboard this handmade craft they called SpaceShipTwo. But the zany ones were often the ones who made history. When Norman Mailer first embarked on his book about the Apollo program, he couldn't make up his mind whether Apollo was "the noblest expression of the Twentieth Century or the quintessential statement of our fundamental insanity."

Branson was not the only one with such ambitions. He had rivals, like Amazon founder Jeff Bezos, with his space company Blue Origin, and Tesla founder Elon Musk, with his company SpaceX. They were all building rockets to take people into space, and Branson was clear that he wanted to be "the first of the three entrepreneurs fighting to put people into space to get there."

Each had distinct visions for the journey. Virgin had pioneered a unique air-launch system—a mothership, WhiteKnightTwo, had been designed to carry SpaceShipTwo to roughly 45,000 feet so the rocket ship would not waste its energy slogging through the dense, lower atmosphere—while others used a more traditional ground-launch system.

Virgin planned to take half a dozen passengers on a "suborbital" flight, cresting about 50 miles above the Earth. By comparison, what is called "low Earth orbit" starts at 100 miles above sea level; the International

Space Station orbits at an average of 150 miles above that; GPS satellites, which operate in "medium" Earth orbit, are about 13,000 miles away.

Blue Origin shared Virgin's suborbital altitude goal for its initial crewed flights but was intent on exploring deep space, too. SpaceX was arguably the most ambitious: Musk wanted to colonize Mars, a minimum of thirty-four million miles away.

But perhaps the most striking distinction boiled down to their belief in the human mind. Blue Origin and SpaceX were run by tech wizards, algorithmic geniuses who trusted in mathematical power to eliminate human error, to one day render fallibility obsolete. Virgin was analog, and despite the futurism of SpaceShipTwo's mission, the vehicle was relatively simple—cables and rods, no autopilot, no automation.

The fate of the ship was in Stucky's hands.

Nichols was sure they were going to die: that was the hazard of crewed spaceflight. "If you want to build confidence in space, don't try sending *people* there," David Cowan, a venture capitalist who has invested in several commercial satellite companies, said. "Any failure will be a catastrophe."

The day was shaping up to be just that. Down on the runway, the lime-colored fire trucks were ready to go. Doug Shane, the president of Scaled, uttered the words that company insiders would recognize as code for a looming disaster. When he said "Blue Zebra," his colleagues knew to plan for the worst.

But Stucky wasn't ready to give up just yet. As the spaceship spun and fell calamitously toward the Earth, he remembered one last thing he wanted to try: he hoped to God it worked.

FOUR YEARS LATER, Stucky and I were sitting around the firepit in his backyard with tumblers of whiskey when he asked whether I wanted to watch the cockpit video from that flight. Coyotes howled in the distance. His wife, Cheryl Agin, drank prosecco and whispered to the two Chihuahuas at her feet.

Of course, I said.

Stucky led me through the house. He was fifty-seven, with a loose-

Mark Stucky, in SpaceShipTwo.

legged stroll, tousled salt-and-pepper hair, and sunken suntanned cheeks. In other settings he could pass for a retired beachcomber. He wore the smirk of someone certain he was having more fun than everybody else.

Plaques and inscribed photos hung from the walls of his home office— "TO THE BEST PILOT I KNOW," FLIGHT INSTRUCTOR OF THE YEAR, "BLUE SKIES AND YIPPEE KAYE." Cartoon stars decorated the ceiling fan. Stucky sat down behind his computer and pulled up the file. It filled the screen.

The video was hard to watch—Stucky straining to avoid passing out; hazard alerts beeping; red lights flashing on the cockpit console; the horizon whipping past.

Agin, his second wife, had slipped into the room and stood over his shoulder, swallowing tears. She had never seen the video.

He hit pause. "Is this emotional for you?" he asked, sounding sharper than he intended.

"It's okay," she replied.

They didn't discuss the dangers of his job much. Death was one of those things test pilots didn't like to ponder. It was "an unpleasant thing to think about," Neil Armstrong said before he went to the moon.

But Stucky didn't sugarcoat it: he was a test pilot for an experimental rocket ship program, a tremendously risky endeavor. Four men had already died for the cause, including Stucky's best friend. "A Marine Corps colonel once told me, 'If you want to be safe, go be a shoe salesman at Sears,'" said Stucky.

Stucky hit Play, and we were back in the middle of the flight, watching him yank and pull on the controls and hearing him strain from the negative g's while the spaceship continued to spin and fall from the sky.

PART ONE

BROTHERS

1.

BORN TO FLY

SHORTLY AFTER MIDNIGHT, John Glenn slipped on a bathrobe to eat filet mignon, scrambled eggs, and jellied toast under a doctor's watchful eye. His blood pressure measured 120 over 80, about the same as when he was home reading a book.

Glenn was nearly six feet tall, with a dusting of freckles and receding, strawberry-blond hair. At forty, he was the oldest member of NASA's first astronaut corps, and, as he said, "there wasn't any closer brotherhood ever formed." He had been plucked from the military and subjected to bowel probes and electrodes and various other assessments to determine his fitness for spaceflight. His high-nutrient, "low-residue" meal of filet mignon and eggs was designed to stop him up.

When he finished zipping up his suit he pulled on a pair of gloves fitted with flashlights in the fingertips to help him see around his six-by-seven-foot capsule in the dark. He rode an elevator nine stories, to the top of the gantry where a crew helped him into the capsule before bolting the hatch shut. He looked through a periscope over the marshlands surrounding Cape Canaveral. A phone line was patched to his wife.

"Don't be scared," he told her, as he squatted atop a 125-ton rocket. "I'm just going down to the corner store to get a pack of gum."

Glenn's folksy earnestness played well in postwar America, the dimpled paperboy who had bundled and sold rhubarb to afford his first bike, gone off and become a war hero, and was now preparing to ride the most powerful rocket NASA had ever built into space. At a press conference he and the other six astronauts had been asked about their family lives, and while the others oozed a certain insouciant masculinity ("What I do is pretty much my business"), Glenn charmed the room.

"I don't think any of us could really go on with something like this if we didn't have pretty good backing at home," he said. And his motivation? "I got on this project because it'd probably be the nearest to heaven I'd ever get."

Shortly before ten a.m., the countdown reached zero and Glenn's rocket rumbled to life. "We're underway," he said. As the engines burned a ton of fuel every second and the vehicle became lighter and the air grew thinner, the spacecraft gathered speed. Glenn felt the g-forces pressing on his chest.

"Little bumpy along about here," he said. He looked outside and saw the sunny Florida skies turning muddy. And then black.

"Zero g and I feel fine," said Glenn.

Down below, where the air was dense and gravity was the law of the land, hundreds of millions of people were huddled around radio sets and tiny televisions and the giant screen unfurled in Grand Central Station. "He's in the hands of the Lord now," said one woman. It was February 1962. The Cold War was on and America was faring poorly. A wall divided Berlin. More than a hundred CIA-trained guerrillas had just been slaughtered on a beach in Cuba.

The Soviets were winning in space, too. First with Sputnik, in 1957, and four years later with Yuri Gagarin, the first human in space. These were massive technical accomplishments with huge national security implications for America: now the Soviet Union could spy on military bases or put a nuclear weapon on the tip of one of their rockets and fire it at the United States. American kids practiced hiding under their desks in case of a nuclear attack. With the fate of the free world hanging in the balance, John Glenn, the first American to try to orbit the Earth, offered a glimmer of salvation.

"Go, baby!" Walter Cronkite, the CBS anchor, cheered on air.

Mark Stucky sat alone, glued to the TV in the den of his parents' home in Salina, Kansas. His mother was doing chores in another part of the house. His father, Paul, was working across the street at the Methodist liberal arts college where he taught physics and astronomy.

Paul had an exacting, scientific mind and a deep curiosity about the universe. He collected fossils and owned a ham radio that he used to send Morse code messages around the world. On clear nights, he would drive Mark and his sisters out among the wheat fields surrounding the city, where buffalo still roamed, so they could gaze up at the sky and call out the constellations.

Mark was just three. But he was old enough to know that something special was happening as Glenn orbited the Earth, all alone in his capsule, speeding more than seventeen thousand miles an hour. Glenn's voice would periodically crackle over the radio, relaying his observations. Like when he sent the rocket booster toppling end over end in his wake. Or when he saw lightning bolts surging through a storm cloud below. Or when "thousands of small, luminous particles" appeared outside the craft.

"Oh! That view is tremendous," Glenn said, at the sight of the Earth's gentle blue radiance.

Mark sat there for hours in front of that wooden TV to see that Glenn made it home. He almost didn't. Chunks of the ship broke off during reentry, and the heat so marred and discolored the outside of the capsule that it nearly burned the painted American flag right off. But there was no denying that Glenn did his country proud: he got dinner at the White House and a ticker-tape parade.

Mark couldn't put his finger on what made Glenn so special—whether it was bravery or patriotism or fame. But when Paul came home that day, Mark told his father that he had made up his mind: he, too, was going to one day become an astronaut.

Paul stared impassively down at Mark. Glenn may have been devoted to his family and faith and country, and been an unimpeachable role model to many. But in Paul's eyes he was a man of unforgivable flaws. Paul saw Glenn's military credentials—the dozens of combat missions he flew over the South Pacific during World War II, the three MiGs he shot down over Korea—not as a testament to his heroism but rather as evidence of his crimes.

Paul had grown up a Mennonite, embracing radical pacifism. Congregants of his church had come to America in the late eighteenth century from central Europe, and still conducted services in German. When World War I broke out, they refused to serve and claimed that war was "a denial of the Christian faith," for which they were ostracized and accused of sympathizing with the enemy; one Mennonite farmer was mobbed and doused in yellow paint while being led to a tree in the town center to be hanged. (He was rescued in time.)

Twenty years later, Paul was a graduate student when Hitler invaded France and Congress implemented another draft. Paul filled out his papers but scrawled "Conscientious Objector" across the form. He spent months in a "civilian public service" camp in Colorado, conducting soil studies and pulling night watchman duty. Later, after Pearl Harbor, he was sent to another camp, in the highlands of central Puerto Rico, where he was interned for more than two years and where he met Mark's mother. "There was never any doubt as to my decision," he said.

Nor had Paul wavered in his convictions since. In the fifties, he moved to Minnesota to work on xenon flashtubes at Honeywell, but left soon thereafter in protest of the company's Pentagon work.

He was now looking down at Mark, as though his son had just proposed some experiment that defied the laws of physics. Paul was dismissive and defiant.

Impossible, he told Mark.

Astronauts came from the military, every one of them, and no son of his was going to serve.

MARK DIDN'T BRING up his astronaut dream again for years, though he privately went on obsessing about it. He perused back issues of *National Geographic* like other boys flipped through nudie mags.

One day, he found an article by an Air Force test pilot titled, "I Fly the X-15, Half Plane, Half Missile." The X-15 was an experimental, air-launched single-seat rocket plane flown by the military and NASA. It was sleek and black, with wings and a tail, and looked like a normal airplane but had a liquid-fuel rocket engine in the back and could go more than six times the speed of sound. "Acceleration from that inferno in the tail pipe pinned me back in my seat," the author wrote.

The X-15 wasn't just fast; the ship summoned enough power to punch through the atmosphere, into space. The author went on, "I was weightless immediately, and it felt pleasant, a welcome relief. The ends of checklist pages on my clipboard rose eerily, and a little cloud of dirt particles drifted up from the floor." Mark was enthralled.

Mark's parents split up when he was eight. One morning, his mom was slicing fruit for his breakfast when she walked out and never came back. Paul remarried, but Mark and his three sisters did not get along with their new stepmother. One time she punished Mark by striking him in the head with a frozen chicken. She did not get along with Paul, either: she once punched him in the nose and broke his glasses; he simply turned and walked away.

Paul opposed the war in Vietnam and brought Mark and his sisters to peace marches. At age twelve, Mark went to see Abbie Hoffman when Hoffman came to Kansas, blowing his nose with an American flag and asking why the "West Point schmuck who plots the Song My massacre" wasn't in jail. An outraged spectator sitting a few seats away from Mark threw eggs at the stage, narrowly missing Hoffman.

Two years later Mark picked up another issue of *National Geographic*; it featured a spread of stunning photographs shot by the Apollo 15 astronauts from the barren, boot-printed surface of the moon.

But Mark was captivated by another article, this one about hang gliding. The author described leaping from the cliffs over Newport Beach, California. "What can I tell you about this first step that encounters nothing solid?" he wrote. "There's nothing to it. This upward stride causes the jaw to drop and the mind to cease its disciplined churning."

Mark convinced Paul to help him buy his first glider. It didn't weigh much but Mark was scrawny, barely a hundred pounds, stumbling around under that eighteen-foot Rogallo-wing kite.

Mark Stucky hang gliding, in 1974.

His stepmom thought he was going to break his neck, but Mark didn't trust her or her risk calculations. He refused to live in fear of death. Life was random and full of accidents: the man who sold Mark his glider was killed on a motorcycle a week later by a drunk driver; two years after John Glenn's big mission Glenn lost his footing in a hotel bathroom, conked his head on the side of the tub, was knocked out and hospitalized.

This was not to say that Mark promptly conquered the sport. He got blown backward and sent cartwheeling down hillsides. But he studied books about flight dynamics and kept marching up that hill overlooking the reservoir west of Salina to try again. Finally, around the time of his sixteenth birthday, he waited for a steady breeze and ran down the hill and felt the kite wanting to lift and took *that first step that encounters nothing solid*. He was airborne. "I was a pilot," he said.

Mark finished high school and enrolled at Kansas State University, in Manhattan, sixty miles east. He majored in physical science but spent most of his time learning to fly. He often skipped lectures and went to the library to study topographical maps of the nearby flint hills. He got really good, constantly exploring ways to go higher and faster. He once mounted a two-stroke engine on his hang glider, buzzing the KSU stadium during halftime of a football game.

One day he convinced some friends to take him skydiving. Mark had recently persuaded Paul to buy him a parachute and, feeling suffused with that unique strain of stupidity that flows from camaraderie, he was eager to try it out.

It was a frigid winter morning when Mark and three friends piled into a single-engine Cessna and took off. Mark sat in the rear right seat and gazed out the window at endless fields of snow, the grain silos like candles on a sheet cake. He was wearing two pairs of long johns under his bell-bottom jeans with a puffy green parka and goggles he stole from the chemistry lab. His floppy brown hair curled around the edges of his helmet.

They flew west toward Salina.

Mark had made arrangements for Paul to pick up his friends and him at the airport and bring them all home for lunch.

As they reached the eastern edge of town, Mark inspected his parachute one last time. It was a backup chute, ill-designed for skydiving with its long suspension lines, small canopy, and no rip cord. But it would have to do.

The pilot, Ralph Fisch, pulled back on the stick and dropped the flaps to slow down.

Mark climbed over the front seat and shimmied out onto the wheel and wing struts. His jacket rippled into the wind. He lowered himself down so his legs dangled like squid tentacles.

Then he let go.

He was free-falling, listening to the whirring sound of the Cessna's propeller fade to nothing and counting a few seconds to himself. He reached for the chute and threw it over his head. *Poof.* The orange-and-white canopy blossomed, easing him down in a snowbank.

Paul was waiting at the airport as planned. He looked confused when he saw the airplane but not Mark. He asked what happened.

"He jumped out," said Fisch.

They piled into Paul's Plymouth Valiant. Fisch led Paul to Mark's pre-arranged landing spot. Mark was standing on the side of the road, Fisch recalled, "this massive billowing orange thing blowing in the wind and this big, shit-eating grin on his face. He looked like D. B. Cooper or something."

Paul scolded Mark and said he bought him the parachute to be safe, not stupid. They never spoke of it again.

When Mark was about to graduate, the alumni magazine profiled him and put him on the cover. BORN TO FLY, read the headline. He was asked about his plans and admitted that he dreamed of becoming an astronaut but, "I hate to tell people that, because it seems like such a kiddie dream."

Still, he took his first step by signing up to join the Marines—just like his hero, John Glenn.

Paul treated Mark's decision as a startling act of defiance. It didn't matter that Mark insisted he was joining just so he could learn to fly. Paul sniffed. He said recruiters made their living by making empty promises and lying to young men.

Mark my words, he added: "In two weeks, you'll be peeling potatoes."

2.

FORGER

STUCKY SURVIVED OFFICER candidate school in Quantico but not the medical exams that followed: his heartbeat was testing lower than the doctors liked. Stucky insisted that it was always slow, but the medics wanted to wait before clearing him for flight school.

Stucky took advantage of the downtime. He flew his hang glider and read a lot of books, including one by Michael Collins, the third and often forgotten member of the Apollo 11 crew. Stucky appreciated Collins's sense of humor; when asked how he felt on the dark side of the moon while Neil Armstrong and Buzz Aldrin were making cosmic history, Collins joked, "I just kept reminding myself that every single component in this spacecraft was provided by the guy who submitted the cheapest tender." The book convinced Stucky that it was possible to be an astronaut without becoming insufferable.

Eventually the doctors agreed that Stucky's heart just beat slow. He reported to Quantico for what the Marines called the Basic School in April 1980. He and some friends drove to DC one night to drink cheap booze at a notorious singles bar with "a meat-market reputation," as a bartender said.

Stucky made eyes with an attractive woman across the room and asked her to dance. Her name was Joan. She was six years older than he was, with a car and a condo in the city. She didn't normally go in for fresh-faced marines. Her father was a JAG officer in the Coast Guard; her mother was close with a rising Army officer, Colin Powell, and his family. She bragged about growing up in the seat of power.

Stucky told her there on the dance floor that he was going to become an astronaut, which she thought was cute. She gave him her phone number and they met for dinner the next weekend. They were married three months later. The wedding was a big deal because she was Black and interracial marriages were extremely rare among Marine aviators, the overwhelming majority of whom were white. Other pilots made comments behind his back. "There was some bigotry in our squadron," said Mark Jean, one of Stucky's squadron mates.

(Frank Petersen Jr., the first Black aviator in the Marines, had been falsely accused of cheating on his entrance exam, kicked off a public bus in training, and arrested for allegedly impersonating a marine officer [himself]; he would describe prevalent and pervasive racism in the Corps after he retired, in 1988, as a three-star general.)

The Marine Corps was a clubby organization, impatient with outliers. When Stucky told another officer about his astronaut dream, the officer advised him not to mention it again: marines were warriors, and the Corps hated yanking its best pilots from combat squadrons. If word got out that Stucky was eyeing the exits, it would harm his prospects for promotion, and he'd never go to test pilot school and he'd never stand a chance of becoming an astronaut.

After flight school he got orders to report to a training squadron in Yuma, Arizona, a hardscrabble town on the Mexican border. Tom Wolfe used to say that flying was "not a craft but a fraternity." Yuma was a fighter pilot's town. They flew and drank and fought together. At a bar one night Stucky was mud-wrestling with a woman in a bikini who cried foul about a frisky hand. The bouncer was threatening to punch Stucky in the nose when the bouncer suddenly found himself surrounded by marines, prepared to defend their own.

Some of the pilots wanted to assign Stucky the call sign "Sumo" after that, but there was a better one in contention. When Stucky got to the squadron he had volunteered to take the visitor badges home because the lamination was peeling and needed some attention. He returned

Mark Stucky at his winging ceremony, in 1982.

with a stack of newly laminated badges, stiff as credit cards. He knew he
didn't strike others as the home ec type; he shared his secret, about how
he and a college friend once ran a sophisticated forging operation out of
his dorm room closet. "Fake IDs, parking passes, football tickets—you
name it," said Tom Berry, the friend.

Stucky's confession earned him the call sign "Forger."

He was more than the sum of his pranks and hijinks: he was an excep-
tional pilot, too. His commanders praised him as an "officer of the highest caliber" with "unlimited career potential" and "superior aeronautical
abilities." "Unquestionably one of the most talented fighter pilots I have
observed in twelve years of flying the F-4," one superior wrote.

Stucky owed some of his success to his remarkable eyesight. "It came
to that, time after time, who could see the farthest," James Salter wrote
in *The Hunters*, his novel about the Korean War. Stucky could spot movement in a distant sky where others saw only clouds. "He routinely picked
up bogeys at like twenty miles out," recalled Jean, Stucky's radar intercept
officer, or RIO.

But he owed much of his success to his wits. Not long after Stucky completed Top Gun, the elite fighter pilot school, he was flying a night patrol mission over the Sea of Japan when he spotted a Soviet bomber in the distance. Stucky caught up with the enemy plane and flipped upside down. Jean, who was sitting in the back seat, looked up and saw the distance shrink between the two airplanes; eventually only a few dozen feet separated the cockpits. Stucky took out a camera and snapped a photograph. When the movie *Top Gun* came out the next year, Tom Cruise's character, "Maverick," performed a similar stunt.

Stucky was thriving in the Marine Corps, but he kept his eyes on NASA. He followed news about the Space Shuttle closely. After twenty-three successful missions, the Shuttle seemed like an orbital juggernaut.

The twenty-fourth was scheduled for January 28, 1986. Shortly after noon, the *Challenger* lifted off.

"Feel that mother go," said the pilot, Michael Smith.

But seconds later a fire burned through one of the boosters, rapidly engulfing the rest of the craft.

"Uh-oh," said Smith, a split second before the Shuttle exploded into a hypergolic fireball, killing him and the other six astronauts on board.

Ronald Reagan gave a somber televised address. "We mourn seven heroes," he said. "They had a hunger to explore the universe and discover its truths."

NASA grounded the program for more than two years while investigators searched for clues, and engineers looked for ways to prevent it from happening again.

Stucky was undeterred. In late 1988, he submitted an application and advanced to the last round but didn't make the final cut. "We regretted having to inform you that you were not selected for the Astronaut Candidate Program," his letter read. "The limited number of openings precluded many highly-qualified individuals, such as yourself, from being selected."

He went to a war in the Middle East instead.

In 1990, after becoming a test pilot, Stucky flew to Bahrain to show the marines there how to use a new piece of software. He flew several sorties into Iraq when the war broke out; he had always wondered how he would fare in combat, whether he would shrink from violence like his father or discover some murderous impulse in himself. He earned a Navy commendation medal for one of his bombing runs. (Later, Stucky's sister Rachel told him, much to Stucky's surprise, that when Paul read about the award in the newspaper he was proud.)

When Stucky got home he applied to NASA again and made the final round, finishing tenth out of seventeen, but NASA took only the top four. "Competition for the program was again extremely keen," an official wrote.

He tried again two years later. Same frustrating result. He wondered if his single-mindedness was setting him up for a life of disappointment. Even people who achieved their dreams sometimes found emptiness on the other side, or realized that the lure of that dream was a mirage.

"I traveled to the moon," said Buzz Aldrin. "But the most significant voyage of my life began when I returned."

NASA DID NOT reject Stucky outright. They liked what they saw and offered Stucky a job as a test pilot. It wasn't exactly what he wanted, but it was something different, and who knew where it might lead. He left the Marines and moved with Joan and his three kids to Houston.

The job was a cruel tease. Stucky flew T-38s, a supersonic training jet, with the astronauts. He knew that he was as good if not better than many of them, but he wasn't one of them. He was like a stagehand who tuned the leadman's guitar.

But he heard their stories about seeing meteors streak through space, and the neon swirls of the aurora borealis, and the pink orbs of plasma that appeared on the Shuttle's surface during reentry—images that simultaneously tickled Stucky's imagination and deepened his disappointment that he would never get to see these sights for himself.

Then, in 1996, a test pilot at NASA's Dryden Flight Research Center had a heart attack, and Stucky got offered his spot. Dryden was at Edwards Air Force Base in Southern California and occupied a mythical place in

Stucky's mind. It was what "test pilot dreams were made of," he wrote, "the premier flight research center in the free world, if not the entire universe." It had been the staging ground for Chuck Yeager's first supersonic flight and the experimental rocket planes that followed. The center was named after a storied NASA engineer, Hugh Dryden, who said the essence of flight test was "to separate the real from the imagined problems."

This credo would guide Stucky's risk calculation for years to come: don't fear the unknown just because it's unknown.

Stucky accepted the job and moved his family again, this time to the high desert. A kind of utility pilot, he flew a range of different airplanes. One mission required outfitting an F-18 with a Cassegrain telescope and a charge-coupled device camera, and then following a pinpoint flight path while an astronomer in the back seat photographed a passing asteroid eclipsing a nearby star. NASA named the asteroid the 13211 Stucky.

Another program set out to assess the viability of air-launching satellites by tethering a rocket ship to the rear of a cargo plane and towing the ship to 30,000 feet and then setting it loose from there—a proof of concept remarkably similar in configuration to the one he would later fly at Scaled. None of the other Dryden pilots wanted any part of this assignment, which was known as the Eclipse Project; cargo planes left a wake of turbulence during takeoff, and no one had ever attempted a towed takeoff behind a cargo plane before. Stucky wound up doing it six times.

"Eclipse is airborne," he declared over the radio.

A superior commended Stucky for his "expertise, diligence, and impressive professionalism," and said his arrival had "greatly increased Dryden's flight operations."

Stucky also got to fly the world's fastest spy jet, the SR-71. Stucky had been obsessed with the SR-71 since he tacked a picture of one onto his dorm room wall in college. He even talked about the airplane in sultry language, describing its "incredible curves," its "tall wide stance like she was about to pounce," and "her skin sinisterly-covered in black." On the day of his flight Stucky wore a yellow pressure suit, climbed to 80,000 feet, and crept past Mach 3, admiring the curvature of the earth and "the infinite azure of the Pacific Ocean."

Stucky picked up random duties, and even some engineering assignments. In 1998, NASA was looking to design a hypersonic research jet, and Stucky was asked to consult with outside experts about the jet's viability. Stucky called Burt Rutan and invited him out for lunch.

Rutan was an eccentric, mutton-chopped engineer and arguably the most influential aerospace designer of his generation; his company was putting out a new cutting-edge prototype every year. One looked like a jumble of toothpicks; it was the first plane to fly around the world without stopping or refueling. Another resembled a prehistoric bird; it set numerous altitude records. "I like to do far-out things with airplanes," Rutan said. Five of his designs would eventually go on display at the National Air and Space Museum.

Rutan agreed to lunch. Before they met, Stucky gave him a general idea in advance of what NASA had in mind. Stucky wasn't expecting more than a brainstorming session. Rutan showed up with sketches.

BURT RUTAN WAS born an engineering whiz. He won so many remote-control airplane competitions as a kid that the hosts changed the rules. In college, his senior thesis won a national award from the American Institute of Aeronautics and Astronautics, and when he went to the institute's award reception in San Francisco, he spotted his idol across the room. It was 1965, Apollo-mania, and there stood Wernher von Braun, the dashing former Nazi who had escaped through the Bavarian forest and surrendered to a US soldier at the end of the war and who was now overseeing the design of a rocket to send Americans to the moon. "He seemed to stand twice as tall as anyone else, and he was surrounded by people who were in awe," recalled Rutan.

Rutan graduated from Cal Poly and went to work at Edwards, armed with his slide rule. He flew with the test pilots and jotted notes from the back seat. Accidents were common; funerals, too. "I was going to the damn chapel every week or two," he said. Rutan was rarely home, and his wife, who took care of their two kids, disagreed with the war in Vietnam and despised Rutan's contributions to it. She finally left him.

He remarried, quit the government, and started his own company in Mojave, a gritty town fifteen miles northwest of Edwards with a history of attracting frontiersmen. Mule trains used to ferry carriage loads of borax from the mines in Death Valley to the nearest railway junction, in Mojave—an ugly burg, "a blot upon fair Nature's face," as one early visitor put it. Lethal pit vipers slithered among the creosote bushes. Vultures hovered by day. Coyotes howled by night. By the early 1900s vigilantes were stalking the railway station, "driving out the tramps like rabbits from the brush" and notching their casualties on "wooden slabs in the municipal cemetery."

In the thirties, the military built an airport to train Navy pilots for World War II. They left after the war, and the runway fell into such disrepair that a local farmer used it to dry his grapes and sell them as raisins. The town saw booms and busts. Investors showed up in the seventies and repaved the runway, and though the town was tough and isolated, the rent was cheap and the sun abundant. It was an ideal spot for flight test.

Rutan built kit planes for a few years, but retail aviation was tiresome and he was increasingly getting offers from private and government clients. He formed a new company in 1982 called Scaled Composites. Scaled became what the *New York Times* called "a creative battering ram for the Pentagon," though its unclassified work earned Rutan global renown— airplanes that flew around the world, a catamaran that won the 1988 America's Cup. It seemed there was nothing he couldn't do. He built a home for himself and his fourth wife and their two parrots in the shape of a white hexagonal pyramid, designed to weather the blistering summers and cold winters, with a waterwheel in the living room for natural humidification. An article in *Air & Space* called Rutan "the magician of Mojave."

Rutan had agreed to meet Stucky for lunch despite his low opinion of NASA—after Apollo, the agency had become a sinkhole for good ideas, he thought. But he was nosy and competitive, always wondering what others were working on and, like Stucky, had space on his mind. He had been thinking about hypersonic aircraft a lot lately—skew-wings and ramjets and scramjets and such. He came to lunch with some sketches for a scramjet that could funnel supersonic air directly into the engine to achieve combustion.

Stucky hurried back to Edwards, brimming with excitement from his meeting and eager to tell his bosses about Rutan's drawings. But they showed little interest. Aviation priorities were changing. Manned missions were out of fashion. Drones were the rage. National security budgets were tight. Stucky soon got fed up and quit.

"It is disappointing to me that the world's premier flight test organization could consider going for extended periods without any . . . piloted research projects," he wrote in a departing note. "Dryden established its reputation by making the impossible possible but increasingly we seem content to make the possible impossible."

Stucky was unsure what he was going to do next.

3.

A BEAUTIFUL BALL

S TUCKY GOT A JOB as a pilot with United Airlines. It came with a spiffy uniform. But when he put the uniform on, the suit sleeves fell short and the shirt cuffs bunched around his wrists.

Still, there was a lot to like. When Stucky flew out to Denver for training, he was met in the concourse near the bronze statue of Elrey Jeppesen, the barnstorming airmail pilot who later flew for United. Stucky was whisked away in a company car. He was impressed. In early 2000, he and Joan and the kids moved into a two-level stucco home with a terra-cotta roof in Camarillo, to be closer to the L.A. airport.

But Stucky had other family concerns. His mother reappeared in his life all these years later, after walking out on him when he was eight. She struggled with mental illness and was obsessed with the Oklahoma City bombing, claiming that she had seen the conspirators on the side of the road earlier that day mixing chemicals in a drum barrel. Stucky tried to disabuse her with established timelines and known facts about the case, but this merely deepened her convictions; she accused her son of covering up for the bombers.

Not long after that she called in a bomb threat. Authorities traced the call and found her in a phone booth on St. Thomas but did not file charges; they put a note in Stucky's file. She may well have ruined his airline career before nineteen Saudi terrorists waving knives and box cutters did so.

September eleventh was a day whose awful and tragic effects would ripple for years. Thousands were killed in the attacks, hundreds of thousands more killed and maimed in the costly wars that followed. The airline industry cratered. United's stock crashed. The company soon filed for bankruptcy.

Stucky came home one day and told Joan that he was about to lose his job.

He had recently joined the Air Force reserves, but the $2,000 a month he got was hardly enough to support a wife and three kids. Money had been tight even before he was laid off. He occasionally brought home sealed airline meals for dinner. Something had to give.

One weekend, Stucky was standing on the sidelines at his son's Pop Warner football game when he got talking with a dad who worked in real estate and who told Stucky he was making "stupid money" brokering mortgages. Did Stucky want a job? Stucky never imagined making house calls or hustling for commissions, but he had a family to feed, so he dusted off his suit and briefcase and got some business cards printed: MARK STUCKY, LOAN OFFICER, PACIFIC REPUBLIC MORTGAGE.

He stopped introducing himself as "Forger."

Stucky was pretty good at the job. He worked the phones and ginned up business through his network of Marine, NASA, United, and hang-gliding friends. But those circles only got him so far. Sales was wild and unpredictable. You ate what you killed: Stucky could put in eight hours and come home with nothing to show for it, or file a mortgage on a single home and make several thousand dollars in commission.

He didn't mind the risk, though. He came up with a scheme. After 9/11, the Federal Reserve was cutting interest rates to boost the economy. Stucky went out and attracted new customers by slashing his fees and offering the lowest available rates, and when his customers were ready to sign, he would lock in the lowest rate.

Or so he said.

But in fact he was holding off on locking in the rate, hoping that it would fall further. *Then* he would lock it in, pocketing the difference. "If the rates would have gone up, I would have had to eat it," he said. "But they always went down."

Flight tests required deliberate and calculated risk management, but this felt like roulette. One hiccup and it would all come crashing down.

"It was riskier than hell," said Stucky.

STUCKY WAS IN bed one Saturday morning when the phone rang. A friend of his from Houston was on the line saying something about the Shuttle being lost. Stucky was groggy and confused.

Lost, he asked?

"Gone," she said.

It was February 1, 2003, two weeks after the Space Shuttle *Columbia* had launched from Cape Canaveral with a cabin full of spiders, silkworms, and bees—specimens for school experiments. On the Shuttle's ascent, a chunk of foam insulation broke off from the external tank and skittered against the front edge of the Shuttle's wing. The Shuttle pressed on, into orbit. Engineers looked into the problem. They found "no safety of flight issue."

There was an open gash on the wing.

Two weeks later, the Shuttle began its descent. As it did, plasma gases seeped into the gash, corroding the wing. Engineers in mission control noticed the air-pressure reading for the tire on the landing gear beneath the damaged wing malfunction. They told the commander.

"Roger," the commander replied.

But it was too late. The weakened wing crumbled under aerodynamic pressure and *Columbia* came apart 200,000 feet over east Texas, falling like confetti.

Stucky knew several of the men and women on board. He flew the missing man formation at the memorial service at Edwards while a bugler played "Taps." The tragedy put his life in perspective.

What was he doing selling mortgages? It could not have felt more vapid and soulless.

America was preparing for another war in Iraq. Stucky volunteered for active duty, emphasizing his "strong tactical background" and graduation from "multiple weapons schools."

The Air Force signed him up.

AS STUCKY WAS getting ready to go to Iraq, a friend called and asked for assistance plucking a space probe out of the sky. NASA had launched the probe a couple of years earlier to gather samples of particles emanating from the sun, known as "solar wind," and which contained traces of krypton and xenon. Now the agency wanted its probe back. It turned to Roy Haggard for help.

Haggard was a prodigious engineer who specialized in midair retrieval, race cars, and high-performance hang gliders. When he was in high school he had designed and built his first kite from bamboo stalks that he chopped down and whittled with his own hands. He met Stucky at the 1974 national hang-gliding championships, and they stayed in touch.

Haggard shared his plan with Stucky over the phone, how the probe would fall out of the sky and deploy its parachute while Haggard or someone on his team would fly alongside it and catch the probe before it hit the ground, to avoid any earthly contaminations; and how once the team caught the probe they would sheathe it and transport it to a secret facility in Utah where the military cultured germs; and how a roof would retract and they would lower the sheathed probe into a clean room.

Haggard said he thought it would work but he needed a partner to rappel from the helicopter to sheathe the probe.

Stucky agreed to give it a try. Two days later he and Haggard each boarded a helicopter to rehearse the retrieval out in the desert. Stucky slid down a rope thrown over the side of his helicopter as it was traveling sixty miles an hour. It was strenuous work and his forearms were screaming but he pulled it off and Haggard told NASA he and Stucky were ready whenever they were.

Stucky was in Iraq when NASA finally called. Stucky had an office on the top floor of a building in a wedding cake palace compound in Baghdad that formerly belonged to Saddam Hussein, with marble bidets and a pond stocked with elephantine bass, and though the country was falling apart, he didn't have much to do. "My biggest threat is a developing case of carpal tunnel syndrome," he wrote.

He was wondering if there was a way for him to rush home, do the thing with Haggard, and get back, when Joan found out that he and Haggard were talking. She objected, forcefully. "Don't start talking shit to Haggard," she said. "*His big concern is his business and cost effectiveness, and using you as cheap labor works for him.* Be realistic about your life/our life. When he calls you allow him to monopolize your time. . . . How long have you known Haggard? How long have you known me? Who is your life with? What's important to you? I ask these questions because I think you forget."

Stucky apologized to Haggard and said he couldn't get back. He followed news about the probe that NASA called Genesis from his office in the palace.

The probe landed in September 2004, like a lawn dart in the middle of the desert. One of Stucky's friends who knew about Stucky's involvement with the probe wrote, "Don't know if you've seen CNN yet today, but the Genesis recovery didn't exactly go as planned. The good news is the probe landed exactly as planned. The bad news is it did it at about Mach 47.2 because the parachute didn't open. I've heard NASA will be blaming you personally."

Stucky called and left a message on Haggard's phone. "I want you to know," he said. "I still would have caught it."

EMAILS HOME TO friends and family captured Stucky's dark, wry sense of humor. Don't expect a lot of pictures, he said, because "I hear that digital photos taken here can sometimes come back and haunt you, even if you don't use a leash"—referring to the crimes at Abu Ghraib.

But his letters also revealed his clear-eyed take on the war. After a harrowing low-altitude helicopter ride over Baghdad, during which the rear gunner tossed handfuls of Starburst candies at kids below, Stucky wrote, "I don't care how much someone might have liked you, if you keep flying over their heads at 100 feet they are gonna eventually get annoyed."

Mark Stucky in Iraq, in 2004.

They were taunting people, he said, "akin to taking a white glove and slapping a gang-banger in the face while insulting his mother."

Stucky didn't venture outside much because he was never sure where the rifle rounds rattling off in the surrounding neighborhoods might land, but sometimes he would go out and look up at what he guessed was the International Space Station.

"It must be kind of strange when you are up there, seeing such a beautiful ball of blue and brown passing beneath you while also realizing you are flying directly over countries consumed with war, strife, famine, and the like—while you sit comfortably, almost antiseptically detached from the rest of mankind," he wrote.

Stucky was halfway through his deployment when he finally heard news that excited him. His lunch partner from Mojave, the irrepressible Burt Rutan, was about to try sending a rocket ship into space on live TV.

Stucky tuned in to watch.

4.

GOD IS HERE

I T ALL STARTED ON A hot St. Louis day in 1996 when Rutan, the head of NASA, and about twenty astronauts gathered near the Gateway Arch to hear an entrepreneur pledge $10 million that he didn't have.

"We are announcing something today called the X Prize," said the entrepreneur, Peter Diamandis. The rules were simple: $10 million for the first team to put a crewed spacecraft sixty-two miles above the Earth. Twice, within two weeks. Without killing the pilot. Before the end of 2004. No governments allowed.

Diamandis modeled the prize on the $25,000 award offered by a New York hotelier in 1919 for the first aviator to fly nonstop from New York to Paris. Four died trying and two disappeared before Charles Lindbergh crossed the Atlantic in the *Spirit of St. Louis* in May 1927.

"Lindbergh changed the way people think about air travel," said Diamandis. "Our goal is to change the way people think about space travel."

A dozen would enter with all varieties of funky designs. Tom Clancy, the novelist, gave $1 million to a team building a rocket-propelled helicopter that looked "more like a Midwestern grain silo that has sprouted a propeller than a spacecraft."

Rutan was favored to win but he spent most of that sweltering day browbeating the NASA head, Daniel Goldin.

"Risks don't register with NASA today," Rutan told him. What happened? How had the organization responsible for Apollo become so hide-bound and sclerotic? It was failing in its mission to inspire. Rutan had some ideas for recapturing that spirit but he was keeping them to himself for now.

"Unless guys like me go out and do this, it will not get done," he said. "Period."

RUTAN HAD THE brains but not the money, so when he looked out his office window one day at the fancy jet taxiing down the Mojave runway, and saw the airstair unfurl, and saw Paul Allen, one of the richest men in the world, descend, Rutan thought, *God is here*.

Allen had flown in to discuss bankrolling an X Prize team, but he wanted to make it clear that he had no interest in sponsoring a suicide mission. He was a software developer, a profession where, as he wrote in his memoir, "Your worst outcome is an error message."

Rutan agreed. He wasn't too worried about getting up, but he was deeply concerned about breaking apart on reentry. In 1967, a friend of his had been flying the X-15 at Mach 5 when he lost control and reentered the atmosphere spinning sideways, a chaotic descent that generated fifteen g's and pulverized the plane in midair.

Allen told Rutan to call him when he knew how to solve the reentry problem.

A year later, Rutan flew to Seattle to see Allen. Rutan was jazzed, ignoring his lunch and the sight of Mount Rainier's spectacular snow-capped peak out the window as he laid out his engineering solution to Allen: movable tail booms.

Rutan explained how, on ascent, the booms would remain fixed and parallel to the base of the fuselage, but when the craft peaked atop its bal-listic path, the booms would swivel perpendicular to the fuselage, trans-forming the vehicle into a shuttlecock and allowing a controlled descent. It was wacky, but as Rutan would say, "A true creative researcher has to have confidence in nonsense."

Rutan called his innovation the "feather." He told Allen, "I would fund it myself if I had the money."

Allen agreed to write a check, on the condition that they proceed in secrecy. Rutan agreed. He worked behind a partition, warding off outsiders by claiming to be busy with a classified project. Rutan liked the intrigue, his broad and wandering intellect naturally tantalized by the unknown.

In his spare time, Rutan read fringe theories about the Kennedy assassination and the Egyptian pyramids. "I do not claim to have solved any of the mysteries," Rutan wrote in the introduction for a retired Army colonel's book, *UFOs: Myths, Conspiracies, and Realities.* He questioned how a hapless drifter like Lee Harvey Oswald could have acted alone, or how Egyptians could have pounded and chiseled those remarkable geometric structures into shape.

Rutan's strident libertarianism fostered his scrappy competitiveness and skepticism of conventional wisdom. He tested aerodynamics without a wind tunnel in college because he was poor, but later, even after he got money from Allen, he rarely saw the need to use one. He sent engineers to the junkyard to look for parts. He tested the feather by throwing foam models off the deck of the control tower. He concocted a thermal protection compound from phenolic resin mixed with red dye and paprika but told people it came from the eyelashes of Nicaraguan racing spiders.

He wallowed in his own eccentricity, though some of his engineers, the select ones involved with the spaceship, couldn't help wondering whether he had gone too far this time, whether Rutan's whims and flourishes had any place in space travel, especially after the *Columbia* crash. Were they being foolish attempting something that NASA and its $1 billion budget couldn't safely achieve?

But Rutan never wavered. He checked his math and checked it again. It was going to work. He *believed.* A rocket ship that he and his team built with their own hands was going to get them to space. He was confident in his own nonsense.

ON OCTOBER 4, 2004, Stucky turned on CNN and sat through an awful series of stories from Iraq about suicide bombings and videotaped beheadings, waiting for the segment about Rutan's spaceship.

Scaled was calling the craft SpaceShipOne, and the mothership, which would carry the spaceship to its launch altitude of about 45,000 feet, White Knight, and now they were on the verge of winning the X Prize. Rutan's best friend, Mike Melvill, had already flown to space twice—one week earlier, and back in June.

The June flight had been historic. Thousands of space enthusiasts showed up. Buses ferried local schoolchildren to the airport. There were tents and RVs everywhere. "Been to two goat ropings and a county fair, but I've never seen anything like this," said Rutan.

When Melvill coasted past the imaginary sixty-two-mile finish line in the sky, he said, "You would not believe the view"; and when Melvill landed he climbed onto the ship and raised a sign that read: "SpaceShipOne, GovernmentZero."

He went up again in late September, but barely made it back alive, as he barrel-rolled into space at two times the speed of sound. "That's a dead guy," said one NASA astronaut who was watching the flight in Houston. But Melvill miraculously steadied the wings and held on. Having tempted fate twice, however, he didn't want to do it a third time. Rutan turned to Brian Binnie instead.

Binnie was not a popular choice, a retired Navy pilot with two Ivy League diplomas at a company where degrees meant nothing and military experience even less. And he was drab: the brown-bag lunch, the head-master's bearing, the rosebud lips. He never quite fit in. And then there was the matter of his flying. On the first rocket-powered test he landed hard and the gear buckled, requiring weeks of repair work. "He flat didn't fly the airplane," Melvill had sniped to a reporter.

Binnie knew what people were saying, how he didn't have the right stuff, and how it was crazy for Rutan to trust him with $10 million on the line. But Binnie swallowed his pride and accepted help from Melvill—a retired Navy commander being tutored by a high school dropout; they practiced dozens of landings until Binnie had it down. On the night before the big flight, Binnie slept on the couch because his in-laws were in town. His mother-in-law spilled coffee on him the next morning.

Thousands of people came to watch. Binnie sat in SpaceShipOne

Mike Melvill (standing) after SpaceShipOne's first successful space flight,
with Paul Allen (left) and Burt Rutan.

while White Knight carried him to altitude. He felt nervous and scared
of screwing up but told himself that his fear was proof that he cared.
White Knight dropped him and he lit the rocket. It was loud and chaotic. The boost felt like a wave that might sweep him away, but when the
rocket burned out the vibrations stopped. "All the tension melts," said
Binnie.

He looked out at the inky blackness beyond his windows, the edge
of the atmosphere a mere ribbon of blue light on the horizon. "Wow, it's
quiet up here," he said.

His body tingled. He had heard people describe space travel as a
search for God. But what if they had it wrong? What if it was in open defiance of Him? A brief triumph over Him. Binnie did not consider himself a
particularly religious man, but he briefly contemplated the Divine. "Surely
if God had wished men to travel into space He would not have created
gravity," Binnie said.

And then he began his descent to Earth, g's building up as thicker air
buffeted the sides of the ship. It felt like driving into a thunderstorm—a

few drops at first, then a downpour, and by the end impossible to see or think in all the mayhem.

He eventually glided down over Mojave, a kind of coiled pattern, like water circling a drain. This technique enabled SpaceShipOne to control its speed using drag and mimicked the landing pattern used by the Space Shuttle and the X-15. Binnie landed flawlessly. "He just frickin' greased it on," said Rutan.

WHEN BINNIE GOT OUT, a broad-shouldered man in a dark suit and sunglasses pulled him aside and said the president of the United States wanted a word. Binnie, Rutan, and Melvill, among others, went into the hangar and huddled around a speakerphone.

"The sky of Mojave is very big. And you've got very big dreams there," said George W. Bush, calling from Air Force One. "Thank you for dreaming the big dream."

Binnie went on *The Late Show with David Letterman*, coming on after one of Letterman's regular guests, Donald Trump.

"How about these people that build a rocket and go into outer space?" Letterman asked Trump. "That sounds like something that would've gotten your attention."

"I would not do it, personally," said Trump.

"That could be the way of the future—space travel," said Letterman.

"Could be," said Trump.

Binnie came out and said Trump was missing the point. "The world has changed," he said. Investors were rushing in. "In three years, people like yourself, Dave. We'll get you a ride up there."

"Is there a beverage cart?"

Binnie was on the road for the next two weeks while his colleagues in Mojave celebrated without him, and they were about ready to get back to work when Binnie returned. He felt depressed and alone, never fully appreciated by his colleagues. What was fame if you couldn't share it with anyone? "There was no closure," he said.

And he wasn't as optimistic about space tourism as his bullish remarks to Letterman had suggested. He couldn't see the business model or the public's appetite for risk. Accidents were inevitable. "No one has a business plan that can gracefully recover from a smoking hole in the ground, particularly if there are casualties," he said. "If the desert is turning red, we're in trouble."

But Stucky was undeterred. SpaceShipOne filled him with elation, envy, and hope. He had given up his dream of becoming an astronaut after all those unsuccessful NASA applications.

Now he wondered if there might be another way.

5.

FOGGLES

IN APRIL 1955, a Beechcraft flew over the tedious wasteland of Nevada, searching for something. When the plane landed and its three passengers stepped out, their feet settled in ankle-deep dust, and they were greeted by a "soul-shattering silence" that made one feel "alone with God."

It was perfect.

One passenger, a top CIA official, returned to Washington and informed the White House that they had found an ideal spot to test top secret spy planes. Construction crews got to work. Three months later pilots and engineers were showing up to a place so buried under layers of classification that they couldn't even agree on its nickname. Some called it the Watertown Strip. Others knew it as "the Site," or "the Ranch."

On the map it was Area 51.

Over time additional runways were paved and structures built. Occupants lived in tiny dorms with creaky floors and sealed windows; they were forbidden from telling their families anything about where they worked or what they did. They commuted in unmarked Boeing 737s that shuttled them between Las Vegas and Area 51, and other nearby secret sites known collectively as "the range."

The fleet of unmarked jets, named JANET, stood for "Joint Air Network for Employee Transportation." Or "Just Another Non-Existent Terminal." Or was the name of the CIA base commander's wife.

No one seemed to know or was willing to say for sure.

On Stucky's fourth Iraq deployment the Air Force offered him a new job. He was in their good graces, having just completed the impossible task of trying to jump-start the Iraqi Air Force. The United Arab Emirates had given the Iraqi government a half dozen turboprops for surveillance and counterinsurgency. The planes were in terrible shape—cables connected with zip ties; holes without grommets; screws and knobs that shook loose mid-flight—and it was up to Stucky to say whether or not any of it was salvageable. Just keeping the stick straight was like "trying to choke a gorilla"; a bullet once pinged the propeller. Stucky flew in body armor, with an AK-47 by his side.

The Pentagon ultimately wrote off the planes, but the Air Force gave Stucky an award for his "incredible resourcefulness and real courage" leading the program.

His new job came with few details, only that he'd be doing top secret flight test on the range.

WHEN STUCKY GOT back to the United States, he packed up Joan and his kids again, and they moved into a two-story stucco home southeast of Las Vegas. He would leave the house on Monday mornings and drive ten minutes to the corner of McCarran Airport reserved for JANET passengers. He would wait in a special terminal with the other top secret commuters until their unmarked Boeing arrived. They would climb a mobile staircase into the jet and disappear for days among the sagebrush and galloping mustangs and the occasional scorpion.

It was bizarre out there. Everyone stayed in their lane. Stucky would run into old friends and want to ask what they were doing there, but he knew such questions were impolitic, and not just nosy but grounds for dismissal. Some visitors had to put on "foggles"—special glasses that obscured one's vision, so you couldn't make out anything you weren't allowed to see.

Stucky got assigned to an F-117 squadron in Tonopah, northwest of Area 51. The F-117 was a treasured national asset and the world's first stealth aircraft. Everything about the planes came under strict secrecy. Even accidents were covered up.

By the time Stucky got to Tonopah, the F-117 had snuck into Panamanian, Iraqi, and Yugoslavian airspace. But lawmakers had concluded that America would be fighting counterinsurgencies forever and mothballed the fleet. That was the official story, at least—that the jets had to be hangared in a way enabling their redeployment, like for a war with China or Russia. It was up to Stucky to make it happen, according to an officer who served with him.

Stucky was used to difficult working environments. But the requirements and restrictions he faced now were altogether unique from the ones he had dealt with in Iraq. Now, as other F-117 pilots have publicly described, he had to contend with enemy satellites looking down on him from space. He would time tests between known satellite overflights so he could take off, hit the test points, and get back down before the next satellite passed. Hardened lean-tos known as "scoot shelters" lined the runway where, in case of unexpected orbital eyes, the planes could temporarily hide; if they were already airborne they would race to a nearby tanker plane and hide underneath until the satellite was gone.

The secrecy strained Stucky's family life. "He would be gone for a week, and couldn't talk about where he'd been or what he was doing," said his son, Dillon, who was in high school. Joan added, "You'd ask him a simple question and he wouldn't answer: he had to go through his memory bank and figure out what lie to tell. He would just stare, silently, like someone hit him with an anal probe." An engineer who had worked on the F-117 in the eighties joked that Soviet intelligence analysts probably knew more about his work than his own wife and children.

Joan felt abandoned. Stucky was off playing spy games, leaving her alone at home. She was used to it, though. For years she had been solely responsible for getting their children to school, football and cheerleading practices, and auditions; all three were actors: Sascha, the oldest, was getting her degree at San Diego State University, paying her sorority dues with income from doing day spots for the TV series *Veronica Mars*. "It was my responsibility to be there for my family and to be a strong base of support," said Joan, a third-generation military spouse. She knew what she was getting into, said Sascha: "She signed up to marry a man with a plan."

What bothered Joan was that Stucky hardly seemed to notice. He paid minimal interest to the daily details of parenting. "He was so self-absorbed that even when he was in the same room he was absent," Joan said. Stucky worked long hours and brought his work home. He would sit with his laptop out while Joan and the kids watched TV or played board games. He missed Sascha's birth because he was in Japan, attending an aerial combat course. Later, when Sascha got older, she rejoiced when her father got picked for jury duty because it meant he would be home for dinner.

Joan demanded that Stucky refrain from making any career decisions on his own. "I need to help you with decisions that affect us. We cannot focus on our life and I cannot support you if your energy and attention are diffused," she said. "As long as I am in your life you need to run stuff by me first."

But Stucky was already thinking about his next move. As a reserve lieutenant colonel without any formal military education, his prospects for promotion were poor. And he abhorred the idea of doing a staff tour in DC. He knew of some civilian test pilot openings, but none of them sounded interesting; he was at the age where employers wanted his experience in the boardroom, not the cockpit.

There was one exception: Burt Rutan's shop in Mojave, Scaled Composites. They were the cool kids in the test pilot world. Stucky knew some of them through his membership in the Society of Experimental Test Pilots, and because Scaled did some classified work, he would sometimes run into their pilots on the range.

But it was a certain unclassified project that had him, and everyone else, buzzing: the spaceship Scaled was building for Richard Branson.

6.

THE REBEL BILLIONAIRE

THE SEX PISTOLS WERE HOT. It was late 1976 and fans couldn't get enough of the band's irreverence. They smashed stuff onstage and sang about anarchy. "We're not into music," said the band's guitarist, Steve Jones. "We're into chaos."

That December, the Sex Pistols went on the prime-time talk show, *Today*. The four glassy-eyed, mussy-haired punk rockers smoked cigarettes and drank booze from paper cups and brought groupies on set. One woman wore a Nazi armband. Another had a starburst painted around her eye.

The host, Bill Grundy, asked the women if they were having fun.

"Always wanted to meet you," said the one with the starburst.

"We'll meet after," Grundy replied, in a lecherous tone.

"You dirty sod," said Jones.

Grundy prodded Jones. "Go on," said Grundy. Vulgarity was good for ratings. "Say something outrageous."

"You dirty fucker," Jones said. The segment came to a swift end, and the Sex Pistols became a rebel phenomenon. Venues canceled upcoming gigs and their record label dropped them.

While most of the music industry recoiled from the band's toxicity, a spunky twenty-six-year-old named Richard Branson saw an opportunity. Branson owned a record store in London and a state-of-the-art recording studio in a stone manor north of Oxford. He had recently issued his first release under Virgin Records, a prog-rock album, *Tubular Bells*. The album sold more than fifteen million copies and made Branson rich.

But the young entrepreneur was terrified of being pigeonholed. Five months after the Grundy debacle, Branson signed the Sex Pistols to his record label.

In May 1977 the band released a single, "God Save the Queen"—a mockery of the Queen's Silver Jubilee that was anything but jubilant; the song bordered on blasphemous, with lyrics like "God save the Queen / the Fascist regime." And the band was getting more outrageous by the day: they fired their bassist for liking the Beatles and replaced him with a junkie named Sid Vicious. Vicious would later admit to killing his girlfriend and was said to have muttered, "I'm a dirty dog," before dying of a heroin overdose.

The BBC refused to play "God Save the Queen," so Branson chartered a boat and set up a stage on deck. They sailed up the Thames, and when they neared Parliament the Sex Pistols broke into "Anarchy in the U.K." and "God Save the Queen." Police came on board and stopped the music.

"God Save the Queen" jumped to number two on the charts.

And the stunt had an even more lasting importance: it solidified Branson's renegade credentials. Those who knew him weren't surprised. Branson had almost been kicked out of one boarding school at age thirteen after the headmaster found Branson in bed with his eighteen-year-old daughter, and Branson informed the headmaster of his next school that he was dropping out to start a magazine.

According to Branson, the second headmaster told him, "You are either going to be a millionaire or are going to jail."

BRANSON NURTURED HIS rebel reputation. "I won't let silly rules stop me," he said. After he made his name in records, he branched out into other sectors.

He fashioned himself as an insurrectionist, proclaiming that Brits didn't have to put up with salty salespeople and crappy goods anymore. "I know that there's need for Virgin to come in and attack a marketplace because I know that I'm frustrated by having to experience bad service in that particular marketplace," he said. Before long, Virgin had its own line of soft drinks, trains, wedding dresses, limousines, wines, airlines, casinos, and condoms.

Recasting tired industries proved to be a masterstroke. It let Branson take credit for reinventions without having to start from scratch every time. Savvy PR spoke louder than novel engineering. Virgin thrived on its brand.

Critics accused Branson of hucksterism. His biographer described him as "a scoundrel, a card player with a weak hand who plays to strength," but also a "self-made and self-deprecating man whose flamboyance endears him to aspiring tycoons, who snap up his books and flock to his lectures to glean the secrets of fortune-hunting." Branson was, in short, a divisive and polarizing figure. He appeared on lists of both British heroes and scoundrels. One poll ranked him second among people whom British children should emulate; Jesus Christ came third.

Branson didn't mind the mixed reviews. He preferred to be in the

Richard Branson, in 1984.

news rather than be ignored, and who could ignore a billionaire standing atop a tank turret in Times Square to promote a soda brand, or dangling naked from a crane with only a cell phone covering his privates as part of its campaign to spread the word that Virgin Mobile had "nothing to hide" on its bills? He flew hot-air balloons across oceans and set a record for the fastest speedboat crossing of the Atlantic.

He survived close calls along the way. During one expedition Branson and his crew had to be rescued when their boat capsized in a nasty storm on the open sea; a Royal Air Force helicopter flew them to dry land. Another time he and his children were attempting to sail a single-hull boat across the Atlantic when their mainsail ripped as they hit a patch of forty-foot waves in the Bermuda Triangle, forcing them to turn back. "We will build another boat and try again," said Branson, ever undeterred.

Branson's appetite for harebrained adventure ultimately led him to Mojave. In 1987, he went there to ask Rutan for advice before one of his ballooning trips. They stayed in touch. In 2002, Branson hired Rutan to build a single-seat airplane that could circumnavigate the world without refueling. A year later, one of Branson's deputies went to check on the project and stumbled onto SpaceShipOne under construction. He immediately called Branson.

"They're building a fucking spaceship!" said the deputy.

Branson had long been fascinated by space travel. He once produced a documentary to commemorate the moon landing, featuring ambient soundscapes, a psychedelic montage of telescope images, and clips of John F. Kennedy's "moon shot" speech. Later, when Branson appeared on the BBC program *Going Live!*, a viewer called in and asked if he'd contemplated any extraterrestrial ventures.

"I'd love to go into space," said Branson. "If you're building a spacecraft, I'd love to come with you."

In 1999, Branson went so far as to register a new company: Virgin Galactic. He liked the name and the idea of a space tourism service. "I hope in five years a reusable rocket will have been developed which can take up to ten people at a time to stay at the Virgin Hotel for two weeks," he once said.

But it was a lark, and when he talked about his pipe dream he sounded more unhinged than prophetic.

YEARS WENT BY. After Rutan won the X Prize in 2004 everyone wanted a piece of him. Pinstriped defense contractors tried to buy him out. Others tried to hire him to do the impossible once again.

The CEO of an electronics company commissioned Rutan to design a propulsion jet for extraterrestrial travel.

Google cofounder Larry Page wanted a flying car to shuttle to and from work, but one that would comply with the Bay Area's strict noise ordinances and fit in a single parallel parking spot. Rutan built a prototype but couldn't make it quiet enough.

Mojave had become a place where the ultrarich could dream of leap-frogging life's otherwise pesky and constraining realities—traffic, gravity. "Nobody complains about making noise or sending plumes into the sky," said one aerospace engineer in Mojave. "This is the Silicon Valley for the new industry."

Branson had his own impossible idea.

He wanted Rutan to build him a rocket for tourists—just like Space-ShipOne, but bigger. Branson had paid a million dollars to Paul Allen to put a Virgin decal on SpaceShipOne, gaining the right to one day adapt the design for tourists. But he couldn't do that without Rutan.

Rutan was tempted, though it felt a little too much like certification and production for his liking. He preferred to spend his energy on novelties. Prototypes. He was an inventor; production was assembly-line work, for lesser minds. But he liked Branson and his rebel billionaire schtick, so he agreed. He saw what Branson saw: the prospects of a ready market.

It wasn't a question of whether people would be willing to pay large sums for a joyride into space: they were already doing it.

FOLLOWING THE COLLAPSE of the Soviet Union, Russia had devolved into a nuclear-armed pawn shop, with anything you desired—minerals, guns, people—available for the right price. Even a seat on a rocket. In 1990, a Japanese TV station paid an estimated $12 million to send one of its reporters into orbit to tape broadcasts and make a documentary about the rain forest. The reporter conducted experiments on Japanese tree frogs.

Others went, too, including Dennis Tito, an L.A.-based investment manager who, in 2001, paid the Russians $20 million for a ride to the International Space Station. Tito, and others who followed, got the full experience—feeling the fuel slosh in the tank during liftoff; the "moments of terror" mixed with "pure joy" that followed; the lonely sound of ticking alarm clocks that filled the cabin after burnout, "like waking up inside the workshop of an old Swiss clockmaker"; the sight of a purple-hued galactic sunset reflecting off the trusses of the space station.

For those with means and the right risk appetite there was nothing else like it; a Himalayan expedition seemed almost pedestrian by comparison.

Moreover, the Russians proved you could charge whatever you wanted. On top of the $12 million, they billed the Japanese TV station $100,000 per hour for the assistance their cosmonauts gave the reporter. These costs proved too much for some: Tito inherited his ticket from an American video game designer who lost his shirt when the dot-com bubble burst; a member of the boy band 'N Sync forfeited his reservation when he couldn't come up with a deposit.

NASA found this all repulsive. When Tito flew to Houston to train with his two Russian companions, NASA officials barred him from entering the facility and later threatened to bill him for disrupting space station operations—though, truthfully, Tito was less disruptive and more meditative, gazing during his trip at the spectacular views out the window and listening to opera on his headphones. He said, upon landing, "I just came back from paradise."

Astronauts had a name for this sensation, the profundity of staring at Earth from space. They called it "the overview effect." One described it as a "feeling of unity." Another said, "You don't see the barriers of color and religion and politics that divide this world." Alan Shepard, the first American in space, said, "It was just very, very emotional." He was moved to tears.

In 1990, when a NASA craft captured an image of the Earth from four billion miles away, framing the planet as a speck awash in cosmic surf, a mere "pale blue dot," the astrophysicist and novelist Carl Sagan said, "That's here. That's home. That's us. On it everyone you love, everyone you know, everyone you ever heard of, every human being who ever was, lived out their lives . . . every hunter and forager, every hero and coward, every creator and destroyer of civilization, every king and peasant, every young couple in love, every mother and father, hopeful child, inventor and explorer, every teacher of morals, every corrupt politician, every 'superstar,' every 'supreme leader,' every saint and sinner in the history of our species lived there—on a mote of dust suspended in a sunbeam."

These accounts made an impression on Branson. He thought weightlessness sounded positively bananas—having to Velcro everything down so it didn't float away, or being able to pitch a dinner roll at your tablemate without worrying that it would fall onto a dirty floor. But what really moved Branson was the transformative power of it all, its almost baptismal nature.

"Once people have gone to space they come back with renewed enthusiasm to try and tackle what is happening on *this* planet," he said. He regarded space travel as a humanistic, rather than escapist, venture. And now, suddenly, it seemed possible that he could offer that experience to the masses.

Or at least to the wealthy ones.

BRANSON PROMISED TO leave the details up to Rutan. Branson was not a micromanager. He had attended his own board meetings for years before learning the difference between net and gross earnings. His skills rested elsewhere, mostly in marketing. He couldn't resist promoting his outrageously cool new venture—even if it meant setting wildly unrealistic expectations.

Space flights would begin soon, he said. "Within five years, Virgin Galactic will have created over three thousand new astronauts, from many countries," Branson announced, in late 2004.

Rutan liked a challenge. But he thought Branson's estimates were way overoptimistic. Scaled had to build and test and license a brand-new spaceship, and they still hadn't agreed on the basic design.

Rutan had been playing with different ideas. He drew a sketch for a ship with room for forty: too heavy. Then he suggested one with room for nine: Branson thought it was better to avoid middle seats. Eventually, Rutan showed up at the Scaled hangar one morning and declared the specs: two pilots, six passengers, twenty-four thousand pounds. Go.

He needed to staff up and fast, but he was choosy about whom he hired. Rutan avoided engineers from the big aerospace contractors because they brought too many bad habits and would "poison the process."

He wanted restless and impressionable minds, engineers ready to prostrate themselves at the feet of iconoclasm, inventors who shared his confidence in nonsense. As he posted in one job listing, "If you are more concerned with interesting, challenging work than in civilization and culture, you might consider a move to our Mojave desert to get in on some real stimulating projects."

7.

HIDDEN TIGERS

LUKE COLBY WONDERED WHETHER IT would be a good fit—him, this ruddy kid from Boston who wore cable-knits and obsessed about rockets; and Scaled, this cowboy company that specialized in airplanes. But when Colby heard that Scaled was hiring a propulsion expert, he flew to Mojave to see what they were all about.

He was met by a couple of senior engineers who showed him around the hangar. SpaceShipOne glistened under klieg lights, radiant with accomplishment. The Smithsonian had recently announced its intent to display the craft in the foyer of the Air and Space Museum, next to Charles Lindbergh's *Spirit of St. Louis* and Chuck Yeager's X-1.

The engineers told Colby that they basically wanted to build a bigger version of the hybrid-fuel motor that they had put in SpaceShipOne. Generally, rockets are classified by fuel. Most rockets use a solid- or liquid-fuel system; both have advantages and disadvantages. Solids are simple but burn like a firecracker and thus aren't ideal for crewed missions: if something goes wrong, you can't turn the rocket off; it is likely to blow up. Liquids can be throttled, but they are exceedingly complex—a Medusan knot of pipes, umbilicals, vents, valves, and cryogenic storage tanks. Inevitably, complexity increases risk.

When designing SpaceShipOne, Rutan thought a hybrid motor would offer the best of both worlds. Hybrids created combustion by combining a solid-fuel component, or "grain," with a liquid oxidizer, so the pilot could shut a valve at any time to halt the combustion process. But the engineers could still mix a solid-fuel grain in a manner that would deliver an even, steady burn.

Colby was relatively new to hybrid motors. He knew solid motors best. He had built Estes kits as a child and was mixing and casting his own solid-fuel grains by the time he was a teenager. Colby's high school mentor told him that people were going to want to call him a rocket scientist but that he should resist because it was a flawed characterization. "Scientists merely study things that *are*," said the mentor. "Engineers build things that can only be dreamed of." It peeved Colby after that whenever he heard someone talk about rocket science, one of those bugbears of propulsion nerdom, like when people called a liquid-fuel rocket a *motor* when it was an *engine*, or called a hybrid-fuel rocket an *engine* when it was a *motor*.

When the interview was over Colby asked the engineers if he could approach SpaceShipOne for a closer look. Usually people first poked their heads through the hatch to see the cockpit. But Colby walked straight to the tail to inspect the motor nozzle. That confirmed his interviewers' hunch: Colby was the right man for the job.

COLBY MOVED OUT west a month later. He traveled light, carrying mostly books. Like Rutan, he had a soft spot for alternative histories and UFOs; he considered science fiction a playground for new ideas.

On his first day Colby was surprised to discover the fledgling and uncertain state of the propulsion program. Scaled had used rubber as the solid-fuel grain on SpaceShipOne, but now they were experimenting with candle wax and it was giving them fits. They desperately needed help, which was overwhelming for someone straight out of grad school. "While I am thrilled to be the one and only 'Rocket Guy' it is a bit terrifying at the same time to not have someone in-house to go to," Colby wrote in a journal.

But Rutan trusted him, and Colby got along particularly well with one of the technicians, a boisterous Southerner named Glen May. May used to zip around Mojave on a rocket-powered bicycle, and he had a rocket-powered canoe that he would tool around the creeks in back home, on the Mississippi-Tennessee border. "I am not just a thrill seeker," said May. "I just have dreams that will not die."

They became fast friends. Colby described himself as a "crazy, hopeless romantic"—he wore a pith helmet and collected antique firearms—and May took Colby around to his favorite spots in the desert, like an abandoned mining town, a volcanic cinder cone, and a canyon where they went and shot Colby's old guns. May's dog Rebel was always close at their heels.

One day, they were showing Virgin's chief operating officer, Alex Tai, around the rocket test site in the desert when Tai's phone rang. It was the company's president calling. Colby heard Tai trying to hurry off the phone; Tai said he couldn't talk because he was getting a tour of the test facilities from the "rocket scientists."

Colby smiled, thinking about his mentor's words. He didn't correct Tai, however. Colby kind of liked the awe and respect he heard in Tai's voice.

COLBY SPENT MORE than a year trying to make the candle wax work. They mixed asphalt into the wax to make it less brittle and more elastic but this made it incredibly toxic. Colby would wear a face shield, mask, and thick rubber gloves to pour the bubbling asphalt into the grain. Sulfurous fumes lingered on his clothes.

In 2006 May's dog died. Colby and May built a small coffin and dug a pit in the hard dirt near the test site that they filled with concrete to fend off the coyotes. Colby honored Rebel by firing an antique cannon.

Both men distracted themselves with work. They felt that they were making history. May thought space travel was a "wide open highway we must travel." Colby agreed; he hoped to "one day take a group of excellent people away from this over-populated, infested rock and start a new better colony somewhere in deep space." He added, in his journal, "Perhaps what we do over the next few years will be inspiration enough to bring us back from the gutter and inspire people to dream again."

On July 26, 2007, Colby arrived at work early. He was preparing to conduct a "cold flow" test later that day. Propulsion is basically an exercise in plumbing, and this test would confirm that the nitrous oxide was properly flowing from the tank to the injector. Colby had recently added a sump tube to suck up nitrous oxide left pooled within the belly of the tank.

Colby did not think this test would be especially risky. By now he and his team were comfortable handing nitrous oxide. According to the scientific literature, nitrous oxide was inert below 550 degrees. Every dentist office had tanks of the stuff stacked in the corner.

That morning Colby drove out to the site and powered up the command post, an old Chevy they called the SCUM truck—Scaled Composites Unit Mobile. Scaled had recently equipped the truck with flat-screen monitors and a mini-fridge stocked with bottled water.

Hours before the test. From left: Todd Ivens, Jim Moran, Luke Colby, Eric Blackwell, Glen May, and Keith Fritsinger.

By midday it was blistering hot, one of those oppressive days in the desert when crow caws echoed for miles and you could picture the ants and the snakes and the prairie dogs huddling under slivers of shade.

A chain-link fence rimmed the test site. Colby walked to the fence to brief his team. There were about a dozen of them standing near the fence. Eric Blackwell, a thirty-eight-year-old tech with expressive eyebrows, and Todd Ivens, a barrel-chested redhead, tried to persuade Colby to stay outside and watch the test with them. Regretfully, Colby said he had to conduct the test back in the SCUM, which was about five hundred feet away, behind a berm.

The others remained at the fence.

Colby initiated the countdown. He opened the red guard over the "ARM" switch and flipped it to "ON," which made high-pressure gas start flowing. When the light beside the "ARM" switch turned red, Colby threw the "FIRE" switch, which sent a torrent of gas into the main valve and sounded like water spilling over the sluice gates in a dam.

But inside the tank something horrible and unforeseen was happening. Scaled's investigation would later determine that the sump tube caught on fire, making the gas expand. The tank was rated to withstand pressures up to eight hundred pounds per square inch; the pressure spiked to sixteen hundred pounds per square inch.

The tank burst.

A shock wave rolled over the desert, broadsiding the SCUM and rocking Colby in his seat. He ran outside, terrified of what he was about to see.

What he saw was worse than he had imagined. The ruptured tank had sprayed millions of jagged pieces everywhere. Shards of flaming carbon rained down. A dust cloud smudged out the sun. Colby hurried to where the fence had stood and saw his friends, bloodied and limbless and crying for help.

Colby bent over Blackwell, but he couldn't detect a pulse so he moved on.

Ivens, the husky redhead, had lost a leg below the knee, and there was blood pouring from his severed stump. But Colby saw fear in his eyes and knew he was alive; he tore off his belt and tied a tourniquet around Ivens's thigh.

Paramedics arrived and pronounced Blackwell dead. They rushed Ivens away in a helicopter, but Ivens died shortly after reaching the hospital.

And there was a third casualty: Glen May. May had been crouching at the fence. He died almost instantly, a few feet from where he and Colby had recently buried May's dog.

As the helicopters took off, Colby, who had given up his shirt and his belt to wounded colleagues, looked around at the wreckage. The fifty-thousand-pound tanker had been knocked over, like some wounded beast, and was venting nitrous in a loud, high pitch.

Colby felt despondent with grief. Back in the hangar, Colby pulled one of the engineers aside and asked if he could stay at his house that night. Colby had dark thoughts swirling around his head and didn't trust himself to be alone.

BRIAN BINNIE HAD WARNED about the desert turning red, and now it had happened. The accident sent Scaled into crisis, puncturing the aura that existed around the company's otherwise impeccable safety record; the line that distinguished bold acts from reckless ones was suddenly hard to decipher. An article said the incident "raised serious questions about safety practices at Scaled." "It was a real blow to our confidence," said one engineer.

Colby went to see a therapist. They met once but Colby didn't go back: he was an engineer and took comfort in analysis rather than reflection. He threw himself into the accident investigation, deciding that the best way to preserve the memories of May and Ivens and Blackwell was to finish the program that they died for.

Rutan was a wreck. Three of his employees were dead, and another three just quit. Rutan's skin turned ashen, he grew weak, and Colby thought he looked like "he could be on his way out." A doctor diagnosed Rutan with constrictive pericarditis, a rare heart disease in which the pericardium, the sac around the heart, thickens and constricts and ultimately chokes the heart like a python. He prepared to take an extended medical leave.

As he was leaving, the company made some changes. They devised an emergency protocol, in case of another accident, one they could trigger by uttering a cryptic phrase that wouldn't mean anything to outsiders but for insiders would be a watchword for catastrophe; they settled on "Blue Zebra."

Also, the company decided it couldn't manage the rocket motor program alone. They needed outside help, "someone to help find all the hidden tigers," Rutan told Colby.

Scaled solicited outside bids and eventually hired a firm called SpaceDev. The decision was one they would soon regret.

8.

EVIL AIR

F UNERALS. STUCKY FELT LIKE he was always coming from one or going to another. They had a sadly predictable rhythm—a grieving spouse; a brotherhood of aviators on hand to pay their respects and say how it could have been any one of them, while convinced that it never would be; and a reading of "High Flight," a poem by John Gillespie Magee Jr., an Anglo-American pilot who flew with the Royal Canadian Air Force during World War II.

Funerals helped Stucky keep up with old friends. He went to one in September 2005 for a NASA test pilot who was performing acrobatics when her cockpit canopy came loose, knocked her out, and caused her to fatally crash. Stucky ran into Doug Shane at the service. Shane was the director of flight test at Scaled, in charge of hiring and assigning test pilots; Stucky and Shane were fellow members of the Society of Experimental Test Pilots, and, incidentally, Shane had joined Stucky and Rutan at lunch that time in Mojave when Stucky was working for NASA. Shane asked Stucky if he was thinking about life after the Air Force and told him to stay in touch.

Shane was cautious. Stucky was a strong candidate, and his résumé spoke for itself. But Brian Binnie had a strong résumé, too, and had never been a great fit at Scaled; Shane was reluctant to hire another military pilot at a company whose DNA was more hobbyist than fighter jock. Then again, Stucky had a good sense of humor and Shane thought he might be different.

They exchanged jokey emails. Shane knew that Stucky worked at Area 51, a source of nearly as much *National Enquirer* copy as purported sightings of Elvis, and Shane told Stucky to say hello to the King. Stucky replied, "The venue out here isn't as large as the big casinos but he appreciates our enthusiasm and enjoys being drug free. He told me to tell you he sends his best."

Meanwhile, Stucky's marriage was falling apart. He was hardly ever around, and when he was home, he preferred to be out with his paraglider, leaping off the cliffsides near Las Vegas. Stucky co-wrote a book titled *Paragliding: A Pilot's Training Manual*, in which he expressed his love for the "elegant simplicity of the sport," how he could imitate the "simple, unencumbered flight of the birds."

Joan complained about Stucky's paraglider taking over their family's life. He would ruin family trips by insisting on bringing the paraglider, she said. "If he's doing that then he needs someone else with him, and it's no fun standing in the dirt while he's doing this singular activity," she said.

They argued endlessly, and though Sascha, their oldest, was by then living in San Diego, she had long since identified what she called "emotional disfunction." "My parents are both children of divorce," said Sascha. "I was giving them relationship books when I was, like, eleven years old, trying to teach them to speak in 'I' statements."

Whereas Joan once found her husband's passion for aviation sexy, she now began to resent him for it. She accused him of having a "sickness" for flight. He believed that she intended this to sound diagnostic. Several of his family members suffered from mental illness. His father had become delusional and paranoid in old age and would say to Stucky, "Tell your friends in the CIA that those window washers aren't fooling anybody." His mother was diagnosed with schizophrenia.

Stucky underwent and passed regular psychological exams in the Air Force. He accused Joan of trying to shame him out of his passion.

Later, when his twenty-three-year-old niece committed suicide, a few years before her twin sister did the same, Stucky spoke at the funeral. He encouraged the bereaved to "hug more often and a little tighter, to *listen* not just *hear*, and to better show our love to those we do."

Because, as he said, "Where there is life, there is hope."

IN MARCH 2008, five months after the Elvis exchange, Shane invited Stucky to Mojave to take a tour of the Scaled facility. Stucky was nervous, eager to make a good impression. Shane took him around so Stucky could see "the truly great things hidden away in those hangars" and let Stucky fly the SpaceShipTwo and WhiteKnightTwo simulator. (WhiteKnightTwo, the mothership, was the unheralded but equally critical piece of Scaled's design, carrying SpaceShipTwo to launch altitude.)

Stucky performed well. Afterward, he wrote to Shane, "You asked me about what I have to offer aside from flying." Stucky knew that Scaled sought more than just handy stick-and-rudder skills in its test pilots.

"I have a reputation for getting things done successfully and one of my strong points is an ability to ferret out the real safety issues from the imagined," said Stucky—the Dryden way, separating the real from the imagined. This wasn't a foreign concept around Scaled, which, Stucky conceded, had an "unequaled ability at figuring out what the real design and flight test issues are and then successfully addressing them." But he wanted to stress that he would fit right in: "I would meld well with your existing philosophy."

A month later, Stucky went paragliding with a guy from the local club, the Desert Skywalkers. They met near a ridge overlooking a dry lake bed south of Las Vegas.

Joan disapproved. She said it would be one thing for Stucky to die doing government work, but another for some hobby. In 1989, David Griggs, a NASA astronaut, crashed and died doing stunts in an antique plane shortly before he was scheduled to launch. A government spokesman cited the "general understanding that when you're named to a prime crew certain activities with heightened risks are off limits." When NASA's Space Mirror Memorial was unveiled two years later with the names of fifteen men and women who had died in the line of duty, Griggs's name was absent.

Stucky knew Joan had a point. He had tried to quit a few times, but he couldn't stay away. He always started back up. He knew the risks. He wasn't careless about them. He didn't have some kind of death wish. But he did prefer, if given the choice, to live a rich life over a long one. His father was in his nineties but miserable.

Stucky looked out from the ridge. Just as he was about to jump he noticed a few dust devils in the distance. "I knew there was some potentially evil air out there," he said. Still, he ran, leapt, and took flight.

He was up for a few minutes, about three hundred feet above the ground, when a funnel cloud blew out his canopy. It sent him spinning out of control. Stucky attempted to accelerate to unstall the wing, but he kept falling. Fast.

He avoided looking down for fear that it would only impair his chance of recovering. "It's kind of like having a revolver and having somebody rushing you and you've got to load the revolver with one bullet: if you sit there and look at him and try to hurry, you're probably not going to do as good of a job as if you're just methodically looking at the bullet, putting it into the revolver, aiming, and then pulling the trigger," he said.

Stucky managed to stop the spinning when he was sixty feet up, but it was too late to arrest his fall. He was too low to throw the reserve chute.

He tensed up and slammed into the dirt.

Stucky lay there a moment, in the fetal position. He was alive but in tremendous pain. His lungs clawed for air and his feet ached and his back felt all kinds of torqued. He crawled fifteen feet to a nearby road and flagged a motorist who called for help.

An ambulance arrived. Paramedics steadied his neck and slid him onto a stretcher. Stucky closed his eyes as they lifted him into the back.

JOHN GLENN ONCE called doctors the "natural enemies of pilots" because seldom did something positive result from an encounter between them. A grumpy quack could declare you unfit to fly because of a toothache or a lazy eye or flat feet. Once Stucky failed to report a concussion when a tree branch fell on his head because he feared the injury could be seen as a liability. Now, as he was going into the CT scan, Stucky feared the worst—that his back was broken, that he would never fly again, that he would be lucky to walk.

The scans confirmed his fears: one shattered vertebra, three compressed vertebrae. The doctor said he was lucky that his back hadn't crumbled like a toy train thrown from a roof. He was also lucky to have this particular doctor, who gave Stucky a choice: he could go in for emergency surgery to fix his back, which would disqualify him from ever flying an ejection seat aircraft again, or he could get fitted for an upper-body cast, stay in bed for a few months, and hope that his back healed on its own. Stucky chose the latter.

He spent the next three months in the body brace, recuperating in the guest bedroom. Sascha was in San Diego. Dillon was about to head off to the Air Force Academy. Lauren, their middle child, was a student at UNLV, living at home and playing nurse.

Stucky's convalescence further stressed his marriage. Joan insisted that he give up paragliding for good. He refused.

"Traumatic injuries are life stressors that can test even the best of marriages and although I had tried to mask it for years, my marriage was never very happy," he later wrote. "My wife had no appreciation for my love for flight." (Joan said that Stucky's characterization of her was "absolutely false," adding, "I supported him in everything he did for twenty-nine years.")

That August, four months after the accident, a hang-gliding friend of Stucky's died of cancer. Joan suggested Stucky stay at home and send a condolence card; he ignored her advice and went to the service.

While in California Stucky called a woman named Cheryl Agin. He and Agin knew each other from Dryden; she had worked in the public-affairs department, where they were friendly, though nothing more; both were married. But they shared an unmistakable attraction. "The pilots were the rock stars, and he was definitely the best-looking of them all," said Agin, a reserved, fit woman with a pixieish haircut. "He would do the coolest flybys"—low-altitude, high-speed passes—"and he would get in trouble for them, but of course we all loved it."

On the phone, Stucky asked Agin if he could come over. She was divorced and living alone in a double-wide in a trailer park forty miles south of Mojave, in Lancaster. Her adult daughter from her previous marriage lived on the other side of town. Agin had worked in DC for a couple of years but hated the place—the people, the traffic, the cost of living. Life was simple in Lancaster, but at least you knew what you were going to get. Nor did she suffer any delusions about Stucky. Agin's father had been a civilian engineer on the X-15 project, so she knew what she was getting into. She accepted that "flying is part of my core," said Stucky.

He returned to Las Vegas and packed his things. In October 2008 Stucky filed for divorce, and moved in with Agin.

He faced swift and severe repercussions. His kids blocked his calls and ignored his emails. When they appeared together in family court, his kids refused to even look his way. Stucky felt his children were determined to erase him from their lives.

Dillon, who had been attending the Air Force Academy to become a pilot like his dad, transferred to UCLA; he made the track-and-field team and competed in the triple jump, but on his university profile page under "personal" he listed only the names of his mom and two sisters. Stucky would travel to his meets but lurk in the bleachers, knowing he was unwelcome and wary of Dillon's rejection.

Stucky wrote a song for Sascha and paid a guitarist to perform it.

> Sascha, oh Sascha, I am so afraid of dying
> While my heart's still a cryin'
> Oh how I long for the day my children will once again be
> My family, my family.

He sent her a recording but never heard back.

Stucky maintained a strict physical therapy regimen with the hope of returning to the cockpit. Six months after he broke his back Stucky was cleared to fly high-g maneuvers in an ejection seat aircraft—a remarkable recovery. He promptly relayed the news to Shane at Scaled. He was hired soon after.

The new job should have been everything he wanted.

But his estrangement from his children was tempering his excitement, and things were only getting worse. A month after Stucky joined Scaled, Sascha wrote a text message on behalf of her siblings, threatening legal action against Stucky if he failed to "discontinue communication."

"You are harassing us," she said. "We will contact your employer and inform them of your volatile and stalker-like behavior toward us."

Her words felt like a "stake in the heart," and he replied in what he promised would be his "final communication" to them.

"You'll never fully fathom the depth of my disappointment," he wrote. "A father attending a public commencement or a conference final track meet does not constitute stalker-like behavior, it constitutes the behavior of a father that wanted to recognize the accomplishments of his children. A father who, when he was younger also competed in sports and also graduated, and deep down in his heart wished his father was in the stands showing his support (he wasn't)."

Stucky had recently attended three funerals and was reminded "just how precious life is and how important friends and family are." He concluded, "In the future, if you find yourself regretting your actions and desire to reestablish contact with me then I encourage you to do so. Love, Dad."

The military had trained him to compartmentalize distractions, but nothing prepared him for how empty and sad he felt. One day he broke his promise and wrote to Dillon, saying, "I divorced my wife, I did not divorce my children," and added, "The only way I can see that I would be so shutout is that you have been convinced that I have wronged your mother and/or am a scumbag. Neither is the case. . . . My previous job required me to sometimes be less than candid about what I did and where I was. That is over. If you will allow it, I would be happy to answer any personal (not prior job-related) questions with 100% truth and then you be the judge and jury of my fate as a father."

No reply. Only a gnawing silence.

Every day, Stucky commuted forty minutes down an empty desert highway to Mojave. But even as he prepared to fly a homebuilt rocket ship, and perhaps even fulfill his lifelong dream of becoming an astronaut, the thing he thought about most was whether he would get invited to his kids' weddings.

9.

THE PARTY

ARK PATTERSON SPOTTED THE GREEN flag and stepped on the gas, his Ferrari growling into the straightaway. He drove up and over a sloping series of turns and went into a hairpin going 120 miles an hour. But he was late on the brakes and he skidded off the track, kicking up a large cloud of dust.

His competitors zipped past.

Patterson wheeled back onto the course, eager to catch up. He was not the surrendering type, and he gradually worked his way back into the race. Patterson had overcome a rare bone disease as a child, one his orthopedist said would confine him to a wheelchair. Ignoring his own dire fate, Patterson stayed on his feet and became an amateur boxer. Later, he made his fortune in private equity. He spent some of it racing Ferraris.

He was advancing through the field when he came around another hairpin. He saw the white flag—one lap to go—just up ahead when another car slammed into the rear of his, pushing Patterson off course. Patterson lost his place on the leaderboard and now had a costly repair bill coming.

Not long after the race Patterson picked up a newspaper and read about Rutan winning the X Prize. It was on the front page of every major paper in the country.

PRIVATE ROCKET SHIP EARNS $10 MILLION IN NEW SPACE RACE, read the *New York Times* headline.

Intrigued, Patterson read more and learned that Branson was promoting a commercial version of the same flight with his new company, Virgin Galactic. The company was catering to a unique client, an adventurous type with plenty of disposable income. Patterson fit the profile.

He followed news about Virgin. Sizing up companies was his "bread and butter." He trusted his instinct for sniffing out frauds but liked what he saw in Branson, with all his spontaneity and derring-do. A space ride sounded hair-raising, and exclusive.

Patterson wanted in.

IN LATE 2004, Patterson applied online for a spot on the spaceship. Weeks passed before he heard anything back. Eventually Stephen Attenborough, the company's slick commercial director, called and apologized for the delay. A flood of applications had crashed the website. Attenborough asked Patterson a bunch of questions about his health. Did he exercise? Any medical conditions? How did he handle risk? Was he afraid of heights? Was he prone to vertigo?

Patterson laughed. He spent his weekdays placing $100 million bets on unstable companies with his investment firm and his weekends on a racetrack. He had just been at Daytona going 190 miles an hour. He thrived under physical and mental stress.

Don't worry about me, said Patterson.

Patterson flew to London and met Attenborough for lunch. They got on well. Attenborough had worked for a prominent investment management firm on the sale side, and Patterson knew the type—big smiles, fake tans, ready to please. Attenborough worked the Davos and yacht show circuits, boasting about "selling a dream" and cracking jokes about how hedge fund sales were different from space trips. Before, he said, "I had to warn clients that products could fall as well as rise. Now I am making that a promise."

Patterson paid $200,000 to reserve his seat. Attenborough told him that Virgin expected to be flying passengers into space within three years. In late 2004 the company said it "envisaged" being "open for business by the beginning of 2005 and . . . operating flights from 2007." This was mostly PR bluster. Truthfully, they had no idea.

There were some legitimate reasons for this. The work was highly technical, and the commercial space industry was new. Plus, at that point Virgin Galactic consisted of just a few salespeople living in London, thousands of miles from Mojave.

But there were legal restrictions, too. Thirty years earlier, lawmakers had passed a bill giving the US Congress authority to scrutinize arms exports. "The flow of sophisticated weapons into areas of possible conflict, without tighter provisions for scrutiny and oversight by Congress, has had the effect of piling up tinder for future conflagrations," said one senator. "Weapons sales are too important to be left in the hands of salesmen." The International Traffic in Arms Regulations legislation, known by its initials, ITAR, prohibited the transfer of "dual-use" technologies, like rockets, to foreigners—or even sharing information with them. Most of Virgin's employees were not US citizens.

"That set the tone between us and Virgin," said Brian Binnie, the SpaceShipOne pilot. Binnie had stayed on at Scaled. He was deeply involved with the early days of the SpaceShipTwo program. Virgin, he said, was "totally unrealistic in terms of setting customer expectations, and it was frustrating for our engineers to see what was being preached and put out in public. We knew what was *really* happening, but couldn't explain the difficulties because of ITAR." Violators had been prosecuted. A pair of US companies had recently paid a $32 million settlement just for helping some Chinese engineers investigate a rocket accident in the nineties. Rutan and his team kept Branson in the dark to avoid exposing themselves to fines.

Therefore, without specifics to convey, the best thing Attenborough could offer was good vibes. "If you are going to be courageous and bring meaningful change, you need to be collaborative," he would say. Others got in on the highfalutin talk. Branson's nephew, Ned Abel Smith, a socialite who would change his name to Ned Rocknroll and marry Kate Winslet, oversaw Virgin's "Cosmic Brand and Marketing" division.

In December 2006 Patterson and his wife flew down to Necker Island. Branson's private escape in the British Virgin Islands, Necker was a tropical paradise with bursting bougainvillea and prancing flamingos, waves for surfing, and reefs for snorkeling. Patterson played tennis with Branson and shared a golf cart with Google's Larry Page. He and his wife attended some other glitzy events until Patterson began to tire of it all. He hadn't paid $200,000 for *this*. When were they going into space?

Constantly, Branson and Attenborough were assuring Patterson and the other customers that success was just around the corner. In June 2007 the company was planning to unveil the design of SpaceShipTwo and WhiteKnightTwo at an event in New York City the coming January. "We will dub the event '2008, The Year of the Spaceship,'" Alex Tai, Virgin's COO, wrote in an email.

A week later the tank blew up.

Richard Branson and Stephen Attenborough, in 2006.

The cold flow accident had rattled everyone. Attenborough went into damage control, hoping to prevent customers from demanding their money back. He assured them that the blast would have a "minimal" impact. Operations would soon be, as he put it, "Business as usual."

Patterson gave Attenborough credit: "He did a magnificent job of puffing up the pillows to make sure that the furniture didn't look sat in."

BURT RUTAN BELIEVED in things. He had made Scaled into a cult as much as a company, forged by his belief in the hazards of perfectionism. He drew his inspiration from another renowned aerospace company, Skunk Works.

Skunk Works was a research-and-development shop for Lockheed, the defense contractor, and had designed some of the most innovative and influential aircraft of the twentieth century, including the U-2, the SR-71, and the F-117. They specialized in prototypes, working fast and without fear of making mistakes. "If we were off in our calculations by a pound or a degree, it didn't particularly concern us. We aimed to achieve a Chevrolet's functional reliability rather than a Mercedes's supposed perfection," said Ben Rich, the former director of Skunk Works.

Rutan parroted this philosophy at Scaled. Engineers who paid too much attention to aesthetics were accused of "gold-plating" their work. When one engineer wanted to design a backlit, push-button switch for SpaceShipTwo, his suggestion was rejected; he was told a basic toggle switch marked with green tape and a Sharpie would do.

Their minimalism applied to safety measures, too. They shied away from building inhibitors and safety latches, instead seeking to hire the best engineers and pilots—finding and mitigating risk through procedures and professionalism rather than hardware. They avoided putting hooded covers on switches and knobs. Some of their prototype designs contained single-point failures. (Critical failure points contained redundancies.) But their engineers were accustomed to working on thin margins. "One bolt falls off and you die," said an engineer.

Rutan and Branson were bound to be an odd couple—one committed to function, the other to flair. And while Rutan admired Branson's passion, he often questioned his judgment. Rutan resisted Branson's plan to put a spaceport in a desolate valley of New Mexico known as the Jornada del Muerto, or Dead Man's Journey: it was twenty-five miles from the nearest town, Truth or Consequences; the views were bad, he contended; and there wasn't a dry lake bed for emergency landings, or even a crosswind runway. But after the state invested $142 million from its budget and raised another $76 million from sales tax, Branson said, "New Mexico will be known around the world as the launchpad for the new space industry." Once they were up and running, Branson added, "We might even be able to allow those aliens who landed at Roswell fifty years ago in a UFO a chance to go home."

In December 2009 Branson threw a party in the desert to show off the spaceship that Scaled had just finished building for him. Royals and actors flew in. So did California governor Arnold Schwarzenegger and his New Mexican counterpart, Bill Richardson.

The party was spectacular and over-the-top, hosted in a twelve-thousand-square-foot plexiglass tent. Above & Beyond, one of the leading DJ acts in the world at the time, provided entertainment. Absolut vodka imported an ice bar from Scandinavia. Scaled engineers got drunk with one of Queen Elizabeth's granddaughters. Luke Colby chatted with Victoria Principal, the former star of *Dallas*.

Patterson drove up from L.A. He didn't drink, and he struggled to make sense of what he was witnessing: Was the party and this whole venture, like Mailer previously said, just an expression of the times? Or proof of its insanity?

Schwarzenegger got up and gave some remarks. Being governor had a lot of perks, he said, but this was "definitely one of the coolest things I have ever done." He ribbed Richardson, a man with a fulsome waistline, noting that Virgin's passengers were expected to experience a few minutes of weightlessness. "There is no one more happy than Governor Richardson about that," said Schwarzenegger.

On cue, trance music pumped out from hulking speaker stacks and floodlights lit up the tarmac, directing everyone's attention to the runway.

WhiteKnightTwo, with SpaceShipTwo suspended underneath, appeared through the fog. Spectators gasped as the airplanes taxied up the runway.

Then the winds picked up.

STUCKY COULD FEEL the wind pushing the aircraft around. A cold front had blown in that day, and the temperature had fallen. But now it was really gusting. Stucky sat in the WhiteKnightTwo cockpit holding the controls, wondering what the hell they were doing. Rain and sleet beat on the windshield. He couldn't believe they were conducting their first mated taxi with SpaceShipTwo in such horrid weather—a violation of all the core tenets of flight test.

Easing the jet into its slip, Stucky could hear the thumping music and see the strobes out of the corner of his eye. When he shut down the engines, an audio clip from the Apollo 11 mission crackled over the speakers: "The Eagle has landed."

Branson came out and met the mated airplanes on the runway. "Isn't that the sexiest spaceship ever?" he asked. Branson urged Schwarzenegger and Richardson to smash bottles of champagne on the nose of SpaceShipTwo, before they all made haste for the warm tent.

Stucky was eager to get the vehicles inside and safe. As he taxied to the Scaled hangar, the winds were blowing increasingly fierce, with gusts above one hundred miles an hour. Several Scaled employees escorted him to the hangar, on the lookout for FOD, or foreign object debris. A squall yanked an antique biplane from its tie-down anchor and sent it skipping across the runway. At one point, Stucky nearly clipped Schwarzenegger's jet with one of WhiteKnightTwo's wingtips.

Branson and his friends were meanwhile huddled around the ice bar, taking shots of vodka.

When Stucky finally arrived at the hangar door the wind was blowing in a sideways swirl of sand and gravel, and now the team faced a further dilemma: WhiteKnightTwo had removable wingtips, and the airplane didn't fit in the hangar with the tips on. By then the wind made it too dangerous to use the scissor lift to remove the tips. Workers filled the plane's fuel tanks with gas to prevent it from blowing away, and erected a wall around it with fire trucks. They waited for the storm to pass.

Back at the party tent, the deputy fire chief worried that the structure was about to collapse. He asked Branson to shut it down.

"Rubbish," said Branson. "There is no way we are going."

The deputy went and commandeered the PA system, ordering an evacuation. Failing to clear the tent within minutes could end in a "major disaster," he said. Buses shuttled Branson and his guests to a nearby hotel. Thirty-six minutes later, the tent roof tore off like the lid of a sardine can, collapsing the walls and scattering debris across the airport.

Shortly after midnight, the WhiteKnightTwo team finally got the wing tips off. They rolled the mothership and the spaceship inside the hangar.

Stucky went to the hotel for a drink but didn't stay long. Some of his colleagues got stuck at the airport when the road home was closed due to black ice. With all of the nearby hotels booked up because of the event, several of them wound up sleeping at the office. Years later, one said, "It's kind of ironic looking back on it, a bunch of the design engineers and shop people that built SS2 stuck sleeping on a couch at work while Branson and his celebrity friends and many of the ticket holders are having a huge party." It was, he said, "a metaphor for the whole program."

Patterson drove back to L.A. that night trying to make sense of the chaos he had just witnessed, because through it all Branson was as confident as ever, telling reporters that SpaceShipTwo would be flying to space in little over a year.

No one from Scaled had any idea where he was coming up with these dates. They still had to put SpaceShipTwo and WhiteKnightTwo through a comprehensive flight-test program that would involve at least a couple of "captive carry" flights (SpaceShipTwo remaining mated to WhiteKnightTwo), a dozen or so "glide" flights (SpaceShipTwo, upon release from WhiteKnightTwo, gliding down for a landing), and several "powered" flights (SpaceShipTwo, upon release from WhiteKnightTwo, igniting its rocket).

And they knew what few others did, that the rocket motor program was in shambles.

10.

ANGEL'S WINGS

LUKE COLBY EYED THE MAN with his finger on the button. Colby was sitting in a bunker at the bottom of a pit used to test rockets. The nozzle of their rocket was squeezed into an enormous sewer pipe that had been fitted into the side of the pit, bent ninety degrees, and run straight up.

It was June 2010. Nearly three years had passed since the nitrous accident. Colby had made a bunch of changes since then, like redesigning the valves that controlled the flow of nitrous from the oxidizer tank into the combustion chamber, installing a burst disk into the oxidizer tank to prevent it from rupturing, and storing the nitrous oxide at a colder temperature.

But Colby felt constrained. SpaceDev, the propulsion company on contract to build the motor, insisted upon minimal outside interference, which made Colby feel like he was working for them rather than the other way around.

At least he still got to sit in the bunker and oversee the tests.

The countdown reached zero, and a SpaceDev tech pushed the button. The earth began to shake. Colby checked the video monitor and saw the plume race through the pipe and shoot into the sky. The bunker shook, and the keyboard letters rattled in their sockets.

Colby was not impressed. SpaceDev was supposed to know what they were doing. Colby winced when he saw the test data: the pressure trace, a graph depicting the range of pressure fluctuation during the burn, was all over the page. The range was nearly three times greater than the desirable limit. Its sputtering was enough, Colby thought, to snap someone's neck.

COLBY HAD JUST started on the job when he first went to Rutan and told the boss that he was doing it wrong, that they were using the wrong type of fuel grain. Scaled had burned rubber on SpaceShipOne, and they were doing it again on SpaceShipTwo. Colby suggested they consider using plastic instead, arguing that plastic would improve performance and burn cleaner.

Rutan liked the idea, a lot. He told Colby that he heard plenty of good ideas every day but very few *great* ones; if this worked like Colby said it would, they might even be able to patent it. Rutan asked Colby to build a subscale, or miniature, motor. Colby did. It ran beautifully. The pressure trace, according to Colby's journal entry, was "so smooth that it looks like someone drew it with a pen."

But that was before the nitrous accident, and before Scaled hired SpaceDev, and before SpaceDev recommended building a rubber-fuel motor, which was the one now shaking on the test stand.

Propulsion was hard. Other privately funded space companies could testify to that. In the late nineties Andrew Beal, having grown exasperated by NASA and the "government-funded boondoggles" hogging the market, formed a rocket company. Beal, an entrepreneur from Dallas who later became a high-stakes poker player, built a test site near Waco. He folded after three years when he realized how much it was going to cost.

Soon after, a prospective buyer came around to the test site. The buyer was fond of wearing a black leather trench coat, displaying the fashion sense of what his biographer called a "high-paid assassin"; had a lot of money from recently selling his online payments company, PayPal; and was now forming a rocket company so he could colonize Mars. His name was Elon Musk.

Musk's early efforts were also fraught. Rockets failed to light. He once burned down a test stand. The noise from tests sent cow herds around Waco into frenzied stampedes. Later he nearly burned down an island in the South Pacific when a SpaceX rocket lifted off, caught fire, turned, and came screaming back to earth. Employees in scuba gear recovered scraps from the surrounding reefs. Every rocketeer "took their lumps along the way," said Musk.

Still, as time went on it became obvious to Scaled and Virgin that nobody was going to space anytime soon with SpaceDev's motor. Virgin's leadership grew increasingly concerned. They were already hedging their bets with alternative rocket options, including Colby's plastic motor and a liquid-fuel engine. Now there was added urgency.

The liquid-fuel option was problematic because of the structural changes required to install it on the spaceship. It was also riskier because liquid engines used complex cryogenics: any additional complexity increased the chance of failure.

Branson was not deterred. In March 2011 he took Musk out for dinner. Musk had come a long way since he was scaring cows and smashing rockets into reefs, but those who knew him were not surprised. A savant, Musk once went out drinking with friends but ended up at the bar, by himself, reading a dusty Soviet rocket manual. He sympathized with Rutan's ethos and criticized government contractors for their indulgence, for "building a Ferrari for every launch when it was possible that a Honda Accord might do the trick," as his biographer, Ashlee Vance, wrote. In 2008, Musk launched the first privately built liquid-fuel rocket into orbit using a two-stage liquid-fuel system. A year later he boosted a commercial satellite into orbit.

Branson told Musk that Virgin wanted to purchase a bunch of rocket engines, along with stage kits, tanks, valves, and batteries. "They wanted a drop-in-place system," said a SpaceX engineer who participated in the discussions. Over dinner Branson made an offer, and afterward Virgin's CEO, George Whitesides, wrote to Musk to say how much he and Branson were "excited to work together to make space history."

Musk wasn't so sure. He didn't exactly need to sell: SpaceX was successful; Musk had recently hosted Barack Obama at one of his facilities. Why should he consider an offer from a competitor? Engineers from Scaled and Virgin kept wanting to talk, but the companies couldn't agree on a price.

Finally, the SpaceX president, Gwynne Shotwell, decreed, "No meeting until the deal is done." They never reached a deal.

STUCKY TRIED TO make do. In October 2010 SpaceShipTwo completed its first glide flight. Stucky's boss and the director of flight operations, Peter Siebold, flew the plane. Siebold, who made the cockpit assignments, had decided that he and Stucky would take turns flying SpaceShipTwo. They spent the next year rotating through the pilot's seat.

A decade younger than Stucky, Siebold was a prodigious engineer whom Rutan had persuaded to quit college in the final quarter of his senior year to come work at Scaled. He was a talented pilot but first and foremost an engineer. Another pilot said, of flying with Siebold, "You could see the spreadsheets written on the back of his helmet." Siebold was a kind of golden boy around Scaled. Young engineers dreamed of following in his footsteps; Siebold and Shane and Mike Alsbury, a thirty-five-year-old engineer–turned–test pilot, with a high forehead and a crooked grin, showed that it was possible to do high-performance flight test without having gone into the military. "Those guys were fucking rock stars," said Elliot Seguin, a Scaled engineer and test pilot.

Stucky was intimidated by Siebold. After Stucky interviewed for the job he had written to Shane and said, "I can point out how to make things better, but, unlike Pete, I can't sit down and write the code to implement them." And yet the more Stucky got to know Siebold, the more he found him to be worrisomely cocky. For aviators, confidence is an asset but arrogance is a liability. As Chuck Yeager, who started his flying career as an ace in the Second World War, said, "Arrogance got more pilots in trouble than faulty equipment."

Stucky once interrupted Siebold in Siebold's office when Siebold was boasting about Scaled's safety record. Stucky didn't dispute the basic facts, how over two decades of flight test Scaled had not suffered a single airborne accident: they flew prototypes without crashing them; they were clearly doing something right. But Stucky was not totally convinced. "Life and death is often quite random in nature," he had written to his friends

Peter Siebold (standing).

from Iraq. "Quite a number of people have told me the soldiers over here are in their prayers. That's great but if prayers were all that we needed then the World Trade Center would still be standing. If prayers were all that we needed us aviators wouldn't need to hold quarterly flight safety meetings. We can make a difference with regards to our longevity but there is still a lot of random fortune and misfortune involved."

He confronted Siebold about the safety record. Stucky said he wondered if the safety data wasn't also a bit deceptive, since it reflected only a relatively small number of flight hours; seven thousand accident-free hours was better than six thousand, but it was still statistically insignificant.

"We could have an accident tomorrow and we would have a *horrible* accident rate," said Stucky. Some pilots went through their whole careers believing that they were immune from human folly—either their own, or the mechanics' who checked the airplane that morning—that bad days were for the guys who didn't have the right stuff.

"It *can* happen to you," said Stucky. "And if you don't think it can happen to you then you are more dangerous than the person who thinks it can happen to them."

IT WAS STUCKY'S turn to fly on September 29, 2011. He and Siebold had been swapping seats for a year, and Stucky was about to fly the sixteenth glide flight. This would be the most aggressive one yet and likely the last one before Scaled installed the rocket motor and began supersonic tests of SpaceShipTwo. It was all part of the process they called "envelope expansion"—battle-testing the spaceship before they could clear it for commercial service.

Expanding the envelope. Pushing the envelope. They were common phrases, references to breaking boundaries and living on the edge. The "envelope" coinage likely originated in 1901, when a Brazilian inventor sailed a silk blimp over Paris. As a contemporaneous piece in *Scientific American* described it, "The balloon is inflated with hydrogen, and in order to maintain at all times a tension on the *envelope*—that is to say, perfect inflation—a compensating balloon filled with air is placed in the interior." During World War II the term gained currency, when pilots and airplane designers referred to the "envelope" as "the engine conditions which will give the maximum economy at any given speed." But the phrase became widely known after the publication of *The Right Stuff* and the movie that followed. In the book Tom Wolfe writes, "The 'envelope' was a flight-test term referring to the limits of a particular aircraft's performance, how tight a turn it could make at such-and-such a speed, and so on. 'Pushing the outside,' probing the outer limits, of the envelope seemed to be the great challenge and satisfaction of flight test."

This particular test was designed to assess SpaceShipTwo's propensity for "flutter," a dangerous phenomenon involving oscillations across the tail and wings that can lead, in extreme cases, to an aircraft breaking apart. The phenomenon was not unique to airplanes: in 1940 the Tacoma Narrows Bridge in Washington state collapsed due to aeroelastic flutter.

The test was a glide flight, and though the word *glide* made these flights sound deceptively tame, Stucky knew better. He would liken high-speed glide flights to three-dimensional chess games—a wise move could open up the board, a wrong move would put him in checkmate. Because unlike a normal flight the one thing Stucky couldn't rely on if he lost control was the one thing that would save him: power.

He strapped in, and WhiteKnightTwo took off and climbed above

48,000 feet. Stucky checked on his copilot, Clint Nichols, and the young flight-test engineer sitting in the back, Wes Persall.

This would be the first flight on which they were taking an engineer on board. Having an engineer on board was not an operational necessity. But there was a publicity event coming up in New Mexico, with Branson scheduled to appear and unveil the new spaceport. Someone thought it would be cool if WhiteKnightTwo and SpaceShipTwo flew in from California for the event. Someone else thought it would be smart to put an engineer in the back, as a kind of mobile mission control center; this flight was to supposed to be a rehearsal.

None of this seemed like such a bad idea until WhiteKnightTwo dropped the spaceship and Stucky pushed the stick forward. That was the moment he realized something was horribly wrong, when he was planning to dive but found himself upside down and spinning out of control, with all of those alerts beeping and lights flashing, trying to figure out how to save the ship. He had pulled on the stick. Nothing. Thumbed the trim switch. Nothing. Tried the emergency trim system. Nothing.

"We're in a left spin," said Nichols.

"Dump, dump, dump," Stucky told Nichols. They were carrying seventeen hundred pounds of water ballast on board to mimic the weight of a rocket motor. Stucky hoped that dumping ballast would shift the center of gravity and steady the ship. He was wrong. And to make matters worse, the dumped ballast immediately froze, caking the windows with ice.

Stucky deployed the speed brakes. Nothing. Stepped on the opposite rudder pedal. Nothing.

He had seemingly tried everything, but he wasn't ready to give up just yet. Stucky had exhausted all the obvious solutions, the ones psychologists referred to as "convergent" ones. But another idea came to him, a distinctly "divergent" one, which neither he nor any of the other test pilots had tried before. Stucky didn't tell Nichols what he was about to do. There was no time. They had to feather. Now.

A decade earlier, when Rutan was trying to get Paul Allen, the Microsoft founder, to fund SpaceShipOne, he had to convince Allen that Scaled wasn't going to kill a pilot in pursuit of the X Prize. To satisfy this require-ment, Rutan had come up with the unique design of moveable tail booms. The booms would rotate, from a horizontal orientation into a vertical one, as the ship prepared to descend; this allowed the booms to bite the air, right the ship, and facilitate a "carefree reentry." Rutan called his innova-tion the feather. Brian Binnie, the SpaceShipOne pilot, likened the feather to a pair of "angel's wings."

Rutan never meant for the feather to be used as a spin recovery mech-anism. But Stucky, disoriented and out of control, figured he had nothing left to lose.

Calmly, Stucky reached for the feather handle, slid it into the "unlock" position, and raised the feather. The vehicle immediately stopped tumbling and flipped right side up. Incredibly, Stucky had regained control. The angel's wings had saved him, Nichols, and Persall from an almost certain death.

Unfortunately he still couldn't see through the thick layer of ice on the windows.

"We're recovered but we're IFR," said Stucky—reliant on instrument flight rules. Wilbur Wright once called flight test a matter of "groping in the dark." He meant it figuratively. But Stucky literally couldn't see a thing. And while recovering from an inverted spin required one set of talents, landing blind would demand another.

Eventually, Stucky found a pinhole-size spot of unfrozen glass, through which he peered, and safely landed the ship. He turned to check on Nichols and Persall: Nichols was white, like he had just seen a ghost. Persall looked equally petrified.

"Welcome to flight test," Stucky told him.

11.

THE GAMMA TURN

Stucky was tiptoeing around the hotel room. He and Agin had checked in the night before, eaten an early dinner, and gone to sleep by eight. Now, at three a.m., Stucky crept into the bathroom to get dressed. He put on his green flight suit. The patch on the shoulder read SPACESHIPTWO, ROCKET POWERED FLIGHT TEST, and displayed three stars, in commemoration of May, Ivens, and Blackwell.

It was April 29, 2013. Siebold had been scheduled to fly that morning but had recently broken his left heel in a paragliding accident. Stucky got his spot—and the chance to pilot SpaceShipTwo's first rocket-powered flight. This was a big deal; the previous flights had been glide flights, but this time they were going to actually light the rocket. Stucky had been dreaming about this day since grade school, when he would memorize passages from that old, dog-eared copy of *National Geographic*, the one with the article by the X-15 pilot describing the "inferno in the tail pipe" that pinned him back in his seat.

Stucky had read accounts from other X-15 pilots more recently, so he could learn what to expect during those sixteen seconds of boost. He knew that a single wayward thought could result in death. He had spent weeks preparing and visualizing every physical and mental step of the flight until those steps felt like the notes of a song stuck in his head. Bravery wasn't what set test pilots and astronauts apart from other humans as much as preparation. "We dissect what it is that is going to scare us and what it is that is a threat to us, and then we practice over and over again so that the natural, irrational fear is neutralized," said Chris Hadfield, a Canadian astronaut.

Stucky stared at himself in the bathroom mirror, his brain recycling that sixteen-second loop. Then he turned off the light and tiptoed back across the dark room. He grabbed a string cheese and a cup of yogurt from the mini-fridge.

He kissed Agin goodbye without waking her. He knew that she had an "overinflated opinion" of him, and he tried not to frighten her. She believed that he was "gifted," and cited one of her friends, a photographer at Dryden, who had flown with Stucky and who told Agin that Stucky was the best pilot she'd ever gone up with. Agin had tried not to worry about him, but he had come home one day recently with documents for her to sign that detailed medical and emergency protocols in the event of a crash. She broke down. "I just had never even thought about the possibility that that would happen," she said.

Stucky quietly shut the door when he left.

THE HOTEL STOOD at the edge of town near a sign that read, WELCOME TO MOJAVE: HOME OF SPACESHIPONE. Stucky passed the sign, hoping others might one day see SpaceShipTwo on a similar sign. Stu Witt, the airport manager, would go to Washington and tell lawmakers, "You can see the future from Mojave." As Stucky approached the hangar he gazed at SpaceShipTwo out on the runway—all lit up and waiting to fly.

Luke Colby, the Bostonian, had already been out on the runway for hours. Colby was the pad manager, a role he assumed after the nitrous accident because he didn't want to be in a bunker again if something happened. As the pad manager, he was in charge of arming the rocket motor. Colby had his team sign the rocket motor; someone wrote FORGER'S FORCE on the side.

At five a.m., Stucky and Colby and the rest of the operations team gathered inside. Colby briefed the group on the latest propulsion specifics, like the temperature and volume of nitrous on board. Then they reviewed the "test card," a chit-sized paper that detailed the guidelines, limitations, and priorities for the flight. Stucky felt sharp and alert as he followed along, even without any coffee, which he shunned on flight mornings to avoid inconvenient bathroom calls.

The test card directed Stucky to burn the rocket motor for sixteen seconds. That was just long enough to break the sound barrier, test the trim system, and check for flutter, though, as they all knew, not nearly long enough to propel them into space.

The limitations of the motor were becoming a source of real concern. SpaceDev was way behind schedule, with little to show for it. Some at Virgin felt they were being played for fools. Frustration was turning to fury. "We need to make some bold moves," said a Virgin propulsion engineer named Jarret Morton, writing to company management. SpaceDev was simply not up to the task. "They are not human spaceflight hardware providers," said Morton. "We should stop paying for crappy ground test motors." His recommendation? "Fire them as soon as possible."

But despite all this gnashing and turmoil in Mojave, the company's sloganeers back in London carried on like everything was fine. Attenborough was out on the road gaslighting anyone who dared to suggest the company was pulling up short.

In 2011: "Delay is a strange word, and there is no delay. Our foot is flat on the gas, we have proven technology . . . I don't think you can ask for a lot more."

A year later: "Things are going incredibly well."

Attenborough touted Virgin's "exciting brand" and its plan to "build momentum with a definite media strategy," because Branson wasn't interested in just creating a business, said Attenborough. It was more fitting to think about Virgin Galactic as a *movement*, as "the democratization of space." He said, "To put poets in space and artists as well as scientists and movie stars and businesspeople? The results of that, I think, will be huge and shouldn't be underestimated."

He didn't talk about risk or the potential perils of rocket flight, about how one astronaut compared a capsule on reentry to "being on the inside of a blast furnace," or how much you would sweat inside the suit. Attenborough only wanted to talk about the euphoria of it all. It was like dwelling on John Glenn's starry-eyed observations from orbit, but ignoring how he was nearly fried to a crisp on his way back to Earth.

Later, when the *Sunday Times* was preparing to publish a story about a problem with WhiteKnightTwo, Attenborough jumped into action. He first tried to dissuade the paper from running the story. When that failed, he emailed customers warning them against taking the article too seriously. It was "predicated on false or deliberately misleading and exaggerated rumors from 'off the record,' nameless contributors," he said.

Everything was going fine, said Attenborough: "As you will know from last week's Future Astronaut Newsletter, things continue to move forward strongly in Mojave thanks to the dedication and skill of our incredible team."

FLIGHT TEST WAS an exercise in creativity and perseverance. Perfection? Not so much. Stucky certainly didn't share Attenborough's rosy assessment. But he was ready to make do with what he had.

After the brief, he and his copilot, Mike Alsbury, the one with the crooked grin, got a ride down to the east end of the main runway where a ground crew had towed WhiteKnightTwo and SpaceShipTwo.

Stucky and Alsbury climbed in and buckled up. Alsbury proceeded to work through a ground checklist. He was studious and humble, rare qualities in a community of cocky pilots. He would sit through briefings, scribbling equations in his notebook, like he trusted his math more than anything he might hear during the brief. He never seemed to "raise his voice or lose his cool," said Brian Binnie, who described Alsbury as being methodical, "like a data computer." Alsbury was the kind of pilot one novelist characterized as apt to "turn any sky into a series of numbers."

A purplish band of light rimmed the horizon. The crew chief closed the hatch.

Colby had been filling the nitrous tank for hours. By this point, said Stucky, "the spaceship goes from being something you could safely park next to at a gas station, to becoming a bomb." Alsbury increased the pressure more, to maximize performance, by flipping a switch that sent helium from a smaller tank, stored in the nose of the ship, into the nitrous tank. The sound reminded Stucky of an old furnace hissing and clattering to life.

WhiteKnightTwo started its engines. Colby, crouching under the spaceship, loved how the vibrations from the jet's spooling turbines felt in his chest. He finished arming the rocket by ratcheting home the final electrical connector. When he confirmed that the voltage was within range, he closed up the fill ports.

"Propulsion system armed," Colby said over the radio. "Have a good flight."

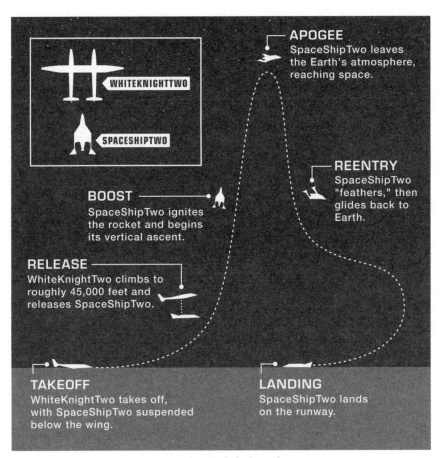

APOGEE
SpaceShipTwo leaves the Earth's atmosphere, reaching space.

WHITEKNIGHTTWO

SPACESHIPTWO

REENTRY
SpaceShipTwo "feathers," then glides back to Earth.

BOOST
SpaceShipTwo ignites the rocket and begins its vertical ascent.

RELEASE
WhiteKnightTwo climbs to roughly 45,000 feet and releases SpaceShipTwo.

TAKEOFF
WhiteKnightTwo takes off, with SpaceShipTwo suspended below the wing.

LANDING
SpaceShipTwo lands on the runway.

SpaceShipTwo's flight path.

He and the other ground crew members stepped off the runway to escape the jet blast as WhiteKnightTwo charged down the runway, took off, and began to climb.

STUCKY MONITORED THE cabin pressure on the way up, while Alsbury worked through the final checklists.

Stucky enjoyed flying with him, and they complemented each other well in the cockpit. They had become close friends while working on a separate, top secret program. Scaled's public projects earned them fame, but the secret ones earned them money. This one entailed a grueling travel schedule. Stucky and Alsbury spent many weeks away from home, in the middle of the Nevada wastelands, where they shared meals and bunked in adjacent, dorm-style rooms.

Stucky felt for Alsbury, who was married with two kids. Stucky knew how much this work could stress a family, and Alsbury had a great one, from what Stucky could tell. He was envious whenever Alsbury talked about his latest family outing—Disneyland, the roller-skating rink, hikes in the mountains where Alsbury carried his young son, Liam, on his back. When Stucky accepted an invitation to throw the opening pitch for the local minor baseball team's Aerospace Aviation Night, he confided to Alsbury that he was worried about embarrassing himself. Alsbury, a baseball fan, offered to help. He met Stucky at the park to practice; Alsbury's daughter, Ainsley, watched from the grass while Stucky threw pitches to her dad.

Stucky was yearning for family. He had taken his ailing father on an Alaskan cruise and traveled many miles to show up at Dillon's track meets, just to watch from a distance. "I've taken flights, rented cars, and even driven all night just to get to watch you or possibly talk to you. Sometimes I didn't speak to you, sometimes you had no idea that I was there, and sometimes you weren't even there," Stucky wrote to Dillon around this time. "Just because I don't scream and shout doesn't mean that I don't hurt or cry."

Now, as WhiteKnightTwo arrived at its target altitude of 47,000 feet,

Stucky activated the release consent switch and waited for the White-KnightTwo pilot to initiate the final countdown.

"Three, two, one: release, release, release."

SpaceShipTwo dropped from the mothership and Alsbury said, "Armed."

"Fire!" said Stucky.

Alsbury flipped the ignition switch.

HIS FLIP OF the switch triggered a synchronized chain of events, each step with its own pyrotechnic logic. Imagine the rocket motor as a shower stall with a hot water tank full of nitrous oxide, a hunk of tire rubber in the basin, and a spark plug hanging beside the shower head.

When Alsbury flipped the switch it sent a powerful current sizzling down a wire that ran from the cockpit to an ignition squib (the spark plug), packed with boron potassium nitrate.

When the current met the squib it showered sparks onto the rubber-fuel grain (the tire).

Simultaneously, a valve between the main oxidizer tank (the hot water tank) and the injector (the shower head) stroked open, releasing thirty gallons per second of nitrous oxide, which sprayed onto the rubber.

When the nitrous oxide mixed with the boron potassium nitrate (the spark plug) and the HTPB rubber (the tire), it set off a controlled explosion in the back of the spaceship, twenty-two feet behind where Stucky and Alsbury were sitting.

A flame shot out of the rear nozzle and sent the rocket ship hurtling forward.

Within about eight seconds, they accelerated to more than seven hundred miles an hour. Stucky's heart rate jumped, and he felt the tingle of adrenaline surging through his body, a tingle to which any honest dope addict, bank robber, or test pilot could testify; their desire to bottle and preserve that tingle was what kept them coming back for more. "Once you experience that, you think, 'Well, I'd like to do that again,'" said the journalist, and recovered addict, David Carr.

But unlike a dope addict sprawled on a couch, Stucky wasn't just along for the ride. He knew that if he savored a moment he could miss a test point. A test pilot was worthless if they weren't hitting their test points.

Three seconds into the flight, a yellow caution light flashed on the console. Alsbury reported, "MOT temp high"—the temperature in the main oxidizer tank, which was filled with nitrous oxide, was running high, and could require them to abort.

"Copy," said Stucky.

He would worry if the light flashed red, but he had other things on his mind, because now they were entering the "transonic zone," and if every test pilot recognized that "for all he knew, there were icebergs of mystery in all he did not know," then the transonic zone was one of those mysterious icebergs.

The transonic zone was a band of airspeed between Mach 0.8 and Mach 1.2. Before Chuck Yeager first broke the sound barrier, many people doubted whether it could be done because strange things happened to aircraft as they were approaching the transonic zone; the doubters worried that some undefined aerodynamic forces would obliterate an object that tried to break through the zone. What kind of strange things? Instrument gauges went wacky. Flight controls stopped working. "It felt like I was driving on bad shock absorbers over uneven paving stones," Yeager recalled of his first transonic experience. In 1956 a Navy test pilot passing through the transonic zone fired his cannons, caught up with his own bullets, and shot himself down.

It was a matter of molecules. During subsonic flight air molecules move to one side or the other of an oncoming wing. But when an aircraft approaches supersonic speeds, the molecules can't move fast enough, and they stack up on the wing, forming shock waves. On Earth, this typically occurs between 650 and 750 miles per hour, depending on altitude, air temperature, and other factors. The shock waves cause transonic instability and make the airplane hard to fly.

"The controls feel like they're in hardening concrete," said Stucky, describing this phenomenon. "We know a lot about supersonic, a lot about subsonic, but transonic? It's the unknown"—the Bermuda Triangle of airspeed. It was, like the island of Rokovoko in *Moby-Dick*, "not down on any map; true places never are."

Alsbury called out their airspeed as they approached Mach 1. All supersonic airplanes experience some kind of transonic instability, twitching and bobbling as they break through the sound barrier. But whereas engineers can program modulations into the flight computers of modern supersonic jets in order to reduce most of the bobbling, SpaceShipTwo was a more austere machine: it had an analog flight controller with circuit cards instead of a digital controller.

The nose bobbled and twitched more than Stucky was expecting. He struggled to keep it level. But these tactile sensations were part of SpaceShipTwo's appeal. And after they rumbled through the unsteady air, the ride smoothed out. They peaked at Mach 1.31.

"There's burnout," Stucky said over the radio as the rocket motor shut down as planned. He set the flight trajectory along a gradual parabolic arc. At the crest of the arc, he fetched a sock monkey from the side console and lobbed it into the air.

"Monkey's in space," he said as the puppet floated briefly, illustrating microgravity for those watching the video feed in mission control.

The flight lasted just thirteen minutes. Branson was waiting for Stucky and Alsbury on the runway when they landed. Agin was there, too, along with Alsbury's wife, Michelle Saling. She and Alsbury had decided not to bring their kids, just in case.

Branson raised his hands to give Stucky a high ten. Stucky liked Branson because he was an adventurer, always ready to push himself to his limits, to test his own mettle. These were qualities Stucky admired. He wished there were more people like him, he said, more people "not satisfied staying within the box, or looking within the horizon." More who wanted to "get on ships and sail out beyond where they can't see, beyond the horizon, into the unknown. That is how the human race advances to the next thing, discovers new worlds, new trade routes," said Stucky. "Voyagers and explorers have always carried more risk than a standard person. But you need the majority to stay behind while explorers go out and establish things and open it up for the masses. And to me that's what space flight is."

They slapped hands. Branson asked Stucky how it was.

Grinning, Stucky said that he'd been tempted to let the rocket burn and head straight for space.

"That's my kind of guy!" said Branson.

THE ENGINEERS KEPT SPACESHIPTWO in the hangar for months after the flight in order to analyze the data and make some repairs: they added a smattering of what looked like miniature fins along the top of each wing to dampen some of the transonic instability.

Stucky felt good about the flight but was wary of a few things, like the tremendous upward forces on the tail of the ship as it passed through the transonic zone. He reminded the other pilots about the importance of keeping the feather locked while they were transonic because if the feather happened to come up while the ship was supersonic it would almost certainly destroy the craft and kill everyone inside.

At this point a more conservative aerospace company might have designed an inhibitor of some sort to prevent premature unlocking. But installing an inhibitor would add complexity, a bugbear at Scaled. This

After SpaceShipTwo's first rocket flight. From Left: Burt Rutan, Mike Alsbury, Mark Stucky, and Richard Branson.

was why the company hired elite test pilots who understood the risks, and the pilots all agreed that it would be wasteful to spend time or money on something so obvious.

Stucky also emphasized the workload—the number of tasks he'd been asked to perform—during boost. It was more consuming than he had expected. He said: "Try lassoing your feet together with a hundred-foot rope behind a horse and then smacking the horse in the ass and having it gallop away while you're trying to do stuff." For future flights he recommended the copilot read out the airspeed when the spaceship hit Mach 0.8, so pilot and copilot knew they were about to go supersonic. He also suggested they tie the sock monkey behind the cockpit with nylon webbing, so they weren't fumbling around with it.

For the next flight Scaled was ready to burn the rocket longer and expand the envelope. But SpaceDev was still struggling to build a stable motor, and no one felt safe riding the current configuration for more than twenty seconds. Four more seconds of burn would barely expand the envelope, but short flights were better than no flights. They scheduled the next one for September 2013. With Siebold still nursing an injured foot, Stucky got the call.

The test card reflected a few variations from the previous flight. Instead of doing another parabolic arc, this time Stucky would pull the nose up and into the "gamma turn"—a steep, near-vertical climb whose trajectory approximated the shape of a hockey stick. And instead of just pushing the nose over and gliding down after the motor burned out, they would deploy the feather and begin their descent floating like a shuttlecock.

For this flight Stucky made less of a production. He slept at home the night before the flight rather than getting a hotel room, and he picked up a hitchhiker on his way into work; his day was already so fraught with risk that he figured bringing a stranded motorist to the nearest gas station was inconsequential. At the hangar, he thanked the receptionist for supplying breakfast burritos, then went about his normal routine.

Movies often construe pilots as mystical cranks, like in *Test Pilot*, when Spencer Tracy's character removes his chewing gum before each flight and leaves it on the tail of his airplane. Stucky had no time for such silliness; he didn't believe in lucky socks or rabbits' feet or lucky breakfasts. He ate string cheese and yogurt on flight mornings because it was easy. "The only superstition I have is not to be superstitious," he said.

He and his copilot, Clint Nichols, posed for a picture on the runway in the dark with an ocher band of twilight framing the horizon. WhiteKnightTwo took off and climbed to its target altitude. The spaceship dropped and Nichols flipped the switch and Stucky said, "Good ignition," while trimming into the gamma turn. Nichols made the "point-eight Mach" call. As they pushed through the transonic zone with ease, Stucky said, "It's sweet here!" The modifications had paid off.

Stucky's and Nichols's wives were waiting for them on the ground, though Branson didn't make it out this time, which often happened on second flights. Alan Shepard was the first American in space and got a ticker-tape parade and an invitation to the White House, while the second man in space, Gus Grissom, got only a lonely room in a shabby guesthouse on base. Stucky didn't mind the lack of fanfare, however, because he saw it as the sign of a maturing test program—that a privately owned rocket ship could break the sound barrier and it would barely register as news.

Besides, the people whose opinions he cared about most took notice. Later that month, Stucky was honored by the Society of Experimental Test Pilots with its Iven C. Kincheloe Award, named after the Air Force pilot who, in 1956, became the first person to fly above 100,000 feet, and in whose name the SETP gave its annual award for the most accomplished test pilot. The award cited Stucky's "flawless" recovery of the near-catastrophic glide flight sixteen and his "significant contributions" to the SpaceShipTwo program.

He was even more pleased when his oldest daughter, Sascha, finally got in touch after several years. Her boyfriend, Jonathan, was about to propose and went to Stucky for his blessing. After Jonathan met Stucky, Sascha decided to give her father a second chance. She made plans to come watch the next SpaceShipTwo flight, scheduled for January 2014.

Around Christmastime Stucky reached out to Dillon to make sure that he knew he was welcome, but under no pressure. "I've got a cool flight test coming up and Sascha's coming over the night prior. We've got plenty of room and a no drama environment," he wrote. "I can arrange/pay for transportation for you if you want."

"Depends on my schedule," Dillon replied. He asked for details, which Stucky shared. Four days later Dillon got back to him. "No way I could find the time right now to head out that way, but I'll keep an eye peeled over the desert," he wrote.

Stucky hoped that he wasn't inadvertently making a mess of things—the tentative embrace from Sascha, another rebuke from Dillon, silence from his other daughter, Lauren. He felt torn and wrote to Dillon, "I hope you and Lauren don't somehow ostracize Sascha since she's established a relationship with me. We all deserve better than that."

On the morning of January 10, 2014, Stucky flew SpaceShipTwo's third rocket-powered flight. Sascha and Jonathan watched from the tarmac as the spaceship pitched into that gamma turn toward the darkening sky. Stucky hugged Sascha when he landed, and five months later she asked him to walk her down the aisle.

STUCKY SHOULD HAVE been overjoyed, with one of his kids back in his life and an enviable career. He'd flown three supersonic envelope-expansion flights in the past year, and if SpaceDev ever got its act together, he might finally achieve his dream of becoming an astronaut.

But he and everyone else now realized that SpaceDev, which had since been bought by a larger aerospace firm, Sierra Nevada Corporation, was floundering and not up to the task. Finally that spring Virgin's CEO, George Whitesides, made the "bold move" that others had been pushing him to make for years—he fired SpaceDev. Virgin was developing a rubber-fuel design of its own, but it wasn't ready, so Luke Colby got the word that his plastic-fuel motor would power the next flight.

Colby was elated. He reflected on everything he had gone through—the optimism he felt when he first arrived, his friendship with Glen May, the horror he witnessed that day at the test site, and the despair and depression that followed. And yet now here he was, having turned the pencil sketch of a plastic rocket motor he had shown Rutan into something that was about to get installed on SpaceShipTwo for its most ambitious flight yet: a thirty-eight-second burn, estimated to go about 140,000 feet above sea level.

Every rocket launch made a statement—from an Estes in the backyard, to a Saturn off the pad in Cape Canaveral, to a SpaceShipTwo flight over Mojave. Each in its own way offered some "promise of collective redemption."

Scaled scheduled the flight for Halloween 2014. Since Siebold's foot was healed, he put himself in the pilot's seat and insisted on Alsbury being his copilot.

Stucky would monitor things from mission control.

The day before, everyone left early to go home and get some sleep. Colby tried to rest but he was all keyed up. He was due back in Mojave at midnight, and he stared at the clock until almost eleven p.m., when he got up and packed his things and commuted into work.

12.

THEY'RE GONE

FINALLY THEY COULD SEE A PATH. SpaceShipTwo had completed thirty glide flights and three rocket-powered flights, and while there was a lot left to learn, the pilots and engineers felt increasingly comfortable with their knowledge of the vehicle's capabilities and limitations, or what they called the "corners" of the envelope.

Virgin's marketing team had moved into high gear, signing a broadcast deal with NBC and securing sponsorships from Land Rover and Grey Goose. The company was firming up plans with Lady Gaga for the star to perform a concert aboard SpaceShipTwo and have the footage beamed live to a stage near Mojave, where tens of thousands of concertgoers would be gathered for a festival. An organizer described plans for the event as "Coachella on steroids."

The moment was further enriched by the tenth anniversary of SpaceShipOne's X Prize flight. On October 4, 2014, Branson attended a ceremony in Mojave to commemorate the historic flight. He said of SpaceShipTwo, "We're on the verge." When someone remarked that Christmas was coming up, and whether Virgin might make it to space before then, Branson said, "I always love Christmas presents, especially if they're early."

Branson flew home to Necker Island after the event. His son, Sam, was on his way to train in a centrifuge outside of Philadelphia. Sam was fit—he did yoga and adventure sports—but he wanted to prepare his body to handle high gravitational forces, or g's; Sam had childhood memories of his dad talking about space travel over take-out curry, but now it was all beginning to feel real. Sam was told that if the next mission, and the two or three that followed, went well, then he and his dad and his sister Holly would be ready to launch into space. "I think any rational person would be slightly nervous," he said.

On the morning of Halloween Branson was home on Necker Island, getting text updates from Mojave.

"Weather trend unfavorable. Will keep you posted," Whitesides told him.

Then, shortly after nine a.m., Whitesides wrote with better news: "Takeoff. So far winds are holding."

ALSBURY AND STUCKY had been in Nevada earlier that week on a top secret job for Scaled. Alsbury never wanted to let anyone down, but when his supervisor discovered that he was scheduled to fly SpaceShipTwo on Friday, he sent Alsbury home to prepare and told him to focus on the spaceship flight.

Alsbury insisted that he could give his full attention to both, before eventually relenting and going home. His selection for the flight was already a source of contention. Virgin wanted to put one of its own pilots, Dave Mackay, a Royal Air Force veteran, in the copilot seat, and since they were paying the bills you might have thought they could force Scaled to accede. But Siebold was the flight-test director, and as long as Scaled was still in charge of the program, no one was going to force Siebold to fly with anyone.

After Alsbury and Siebold flew the simulator on the day before the flight, Alsbury went home early. He and his wife, Michelle Saling, attended their daughter's dance practice and then went to Family Night at their son's school. Their son was in second grade and wanted to become a pilot. They ate early, and Alsbury said good night to his kids and was asleep at nine.

He got to the hangar the next morning around four a.m.

Stucky, who would be assisting the test conductor in mission control, arrived about the same time. He didn't like what he saw when he reached the hangar. Things felt a bit disorderly: a sensor system in SpaceShipTwo had fuzzed out and was being replaced; the nitrous oxide truck had come late, so the liquid inside was too cold—thermal blankets were now draped over the tank.

Siebold was off pacing the tarmac in the dark.

From left: Clint Nichols, Mike Alsbury, Brian Maisler, Mark Stucky, Dave Mackay.

LUKE COLBY'S HEART was hammering. He didn't know if he had ever felt so nervous. This was *his* rocket motor, and he was ready to own every step of the test. He had been at the wheel of the front-end loader earlier that week when they jacked the motor into the air and slid it flush into the back of SpaceShipTwo, and he had been out there on the runway the whole night before, overseeing the pressurization of the rocket motor.

Colby knew that this might well be his last test flight. He hadn't told many people but he felt like he had to get out of Mojave; whenever he left town and went someplace temperate and green his mood immediately lifted, which made him realize that Mojave was making him depressed.

He had recently gone back to Boston and decided that it was time to move home. He signed a lease on an apartment in the Back Bay. He planned to set up his own boutique propulsion shop.

By eight a.m. the sensor system had been fixed and the nitrous temperature had risen. The mission was on.

Saling arrived at the airport with her kids, Ainsley and Liam. She and Alsbury had decided to let them come because Alsbury didn't know whether he would have a chance to fly SpaceShipTwo again. He wanted his kids to witness him making history. It would be their first time watching him fly the spaceship.

A van drove Alsbury and Siebold to the end of the runway. They shouldered their parachutes and stepped into SpaceShipTwo. The sock monkey was knotted behind them.

Colby was crouching under the wing waiting for clearance to ratchet the arming circuit. He realized how much he was going to miss all of this. When he got clearance he closed the panel over the dump port and cleared the pilots for takeoff.

"Propulsion systems armed," he said. "Have a good flight."

WhiteKnightTwo and SpaceShipTwo zoomed down the runway. The sight brought Colby great joy and relief. He knew the amount of effort and sacrifice that had gone into getting them here. But there was not time to revel: he hopped into the van and hurried back to mission control so he could monitor the propulsion data in real time.

THE MATED PAIR flew about 150 miles northeast, alongside a craggy mountain range streaked with purple and green mineral deposits. After reaching Death Valley, and still ascending, WhiteKnightTwo, with Space-ShipTwo attached like a marsupial to its belly, looped back toward the release point. Siebold and Alsbury flooded their masks with oxygen in case of an emergency. They ran through their checklists.

Alsbury: "Seat belts and shoulder harnesses?"

Siebold: "Snug."

Alsbury: "Rocket burn timer?"

Siebold: "Set and verified."

Alsbury: "Stick."

Siebold: "Stick is forward."

Siebold went through the bailout procedure in his mind, feeling for the d-ring of his parachute and the buckles of his harness. Neither was wearing a pressure suit because the design of the ship was sufficiently redundant to preclude the need for a suit.

Alsbury armed the release switch, prompting the countdown: "Three, two, one: release, release, release." The spaceship dropped from White-KnightTwo.

Alsbury fired the rocket.

"Good light," said Siebold, his voice reedy from the onset of g-forces.

Down in mission control, Colby glanced at the incoming data. He wasn't surprised to hear Siebold straining because his motor was producing about 20 percent more thrust than the other motor had on the first three powered tests. Colby stared at a pair of split screens—one showing data, the other showing the rocket plume from a camera mounted on the tail boom. He had never seen or felt anything like this before. The motor was running better than he could have dreamed—plastic burned cleaner than rubber, so the plume was almost translucent, with shades of orange at the edges and a beautiful Mach diamond in the middle.

It's working, Colby was saying to himself.

STUCKY WAS ACROSS the room, monitoring the same feeds. Following Stucky's advice, Alsbury called out the airspeed when the spaceship hit Mach 0.8. They were in the transonic zone. Everything was going well.

"Yeehaw!" said Siebold.

But then Alsbury did something inexplicable. He reached for the lever that controlled the locks on the feather. "Unlocking," he said.

Stucky's body seized. Had he misheard Alsbury? Or failed to notice that the spaceship had passed through the transonic zone? Stucky checked the Machmeter on the main display screen. No: the spaceship was still below Mach 1. Without the locks in place, Stucky knew the upward aerodynamic forces would push the feather up and create a tremendous amount of drag, shredding the ship apart in midair.

Colby watched the vehicle on his screen begin folding in half.

Stucky lunged at the call button to scream *Don't* but he knew it was already too late.

Siebold grunted in agony as the g-forces spiked and when the audio feed stopped and the video froze mid-frame it showed Siebold and Alsbury thrust forward in their seats with their heads nearly in their own laps.

An engineer looked up from his console and gave Stucky a searching look.

"They're gone," said Stucky.

13.

WRECKAGE

SIEBOLD HEARD A LOUD BANG followed by what sounded like "paper fluttering in the wind." He was still strapped in his seat but the spaceship was coming apart around him—the wings and tail booms ripped off, the rocket motor followed. Where the rear cabin had once been was now an infinite blue sky. Siebold blacked out.

He was about twenty-five miles north of Mojave and 45,000 feet above the ground. This was dangerously high for someone not wearing a pressure suit; oxygen was scarce, and the temperature was about negative sixty degrees Fahrenheit. Pilots had survived exposure at such altitudes, but their experiences were unpleasant. In the late fifties an F-8 pilot ejected into a thunderstorm at 40,000 feet and lived, but when he landed his ears, nose, and mouth were oozing blood. In the mid-seventies an SR-71 pilot bailed out at 79,000 feet, but he had on a pressure suit that saved his life.

Siebold had somehow wriggled out of his seat and was falling like a rock. He stirred awake to find his helmet cockeyed and twisted. His eyes burned; he couldn't open his right one, but through his left he saw the ground below coming up fast. His parachute puffed and snapped him fully awake.

Only now as Siebold gasped for oxygen and floated beneath that bright orange canopy could he begin to fathom what had occurred. He was cut and bleeding. His right shoulder dangled from its socket, and he feared

Wreckage, north of Mojave.

that his back was broken, but at least he was alive. He landed in a creosote bush near a dry lake bed and waited for help.

STUCKY RAN TO the roof to get a better look, but streaky clouds to the north obscured his view. He rushed back to the control room.

A manager picked up the phone and declared: "Blue Zebra."

Stucky pulled the Blue Zebra binder off the shelf and flipped through pages of procedures, maps, hazardous material data sheets, and contact information for emergency crews and state agencies. He dialed 911.

Michelle Saling was on the tarmac with her children. She knew she had to get them out of there before whatever horrific fate was about to reveal itself. A friend offered to take the kids home.

Saling and Siebold's wife, Traci, were ushered into a makeshift command center in the Scaled hangar. Both women feared the worst. They were told something almost more unbearable: that one of their husbands was alive, but the other was not. They could grieve together for a moment, but soon one of them would be counting her blessings while the other was left alone in the world; Saling fought the urge to wish doom on another human.

Stucky hopped into a small plane and flew to the crash site.

EMTs had already arrived. They braced Siebold's neck and were administering oxygen. And though his collarbone was broken and his right arm was bleeding through his flight suit, he was going to live. A helicopter transported him to a nearby hospital, where a doctor removed pieces of fiberglass from his eye.

As the plane circled overhead, Stucky scanned the desert for wreckage. Crews found the cockpit twenty miles north of Mojave. It made a nine-foot-wide, three-foot-deep crater in the dirt beside the highway. Officials cordoned off the area with police tape, but the wreckage spread far and wide. A FedEx driver narrowly avoided one piece of falling debris, and what he, a military veteran, saw when he got out of his car reminded him of Iraq. "It looks like the scene out of a car bomb or a roadside bomb," he said. "Parts everywhere, and unfortunately there's pieces of the guy everywhere, too."

Stucky returned to the site in his SUV and joined the search. The desert floor looked like a crime scene. An engineer pointed Stucky toward the shrubs, at what appeared to be a human bust. Immediately, Stucky knew. He went over and saw Alsbury's badge around his neck. Stucky wanted to take the badge off but resisted the urge because doing so somehow felt like an invasion of his friend's privacy.

IT TOOK LUKE Colby a moment to process what happened. One second he was watching the monitors, admiring the steady pressure trace and the hue of the rocket plume. The next he was staring at a fuzzy screen and feeling his delight curdle into despair, thinking, *Not again.*

He knew that rocket better than anyone, so he jumped into a truck with another engineer and sped to the site to look for signs of chemical spillage or other hazards. The wreckage told a story. When Colby saw the nitrous tank among the debris, still fully intact, he knew it vindicated their decision to stick with the hybrid motor; had they switched to a liquid engine, the propellants would have probably mixed in midair and created a giant fireball, leaving no survivors. They were lucky, in that sense.

Because of the altitude and speed at which SpaceShipTwo was traveling, pieces landed across an area dozens of miles wide. One piece floated down like a maple leaf onto a golf course green fifty miles away. Most of it landed in less welcoming parts. This was hard country. Kern County was one of the largest counties in the United States. People moved there for a reason. They wanted to be left alone to do their thing. Some cooked meth. Others plotted against the state. When police got a tip that a miner was preparing to blow up the courthouse and go "hunting" for cops, they raided his place and found booby traps and bombs.

Siebold landed a few hundred yards from the front door of one house. No one came out to offer help. At another residence a sheriff's deputy was greeted by a man in a bathrobe holding a shotgun and surrounded by dogs.

Feds in nylon windbreakers fanned out. Investigators from the National Transportation Safety Board picked through the wreckage, looking for answers. They hauled pieces back to a hangar in Mojave for further analysis. They conducted interviews and reviewed the flight data and watched, and rewatched, feeds from the cameras placed inside the cockpit and outside the ship. "Because it was a test flight, it was heavily documented in ways that we don't usually see in normal accidents," said Christopher Hart, the NTSB's acting chairman.

But Stucky knew that there was no great mystery. His friend made a simple and fatal error: Alsbury prematurely unlocked the feather, which exposed the spaceship to an onslaught of aerodynamic forces in that tumultuous transonic zone. Investigators could spend a lifetime trying to uncover the mystery of why he made that unforgivable mistake. Was it stress? A design flaw? Inadequate training? Stucky couldn't fathom what would make Alsbury "skip a chapter in the hymnal," as he put it. And no one would ever know for sure.

Stucky returned to the crash site nearly every day for weeks. Supposedly he went looking for wreckage. Some parts hadn't been recovered. Agin, who went with Stucky on the weekends, said, "We didn't want pieces of the spaceship to end up on eBay." They searched in vain for the missing sock monkey, though other things gradually turned up.

One day an anonymous caller phoned Scaled's front desk and provided geocoordinates for where they might find some missing stuff; the caller said one of his friends had been stalking coyotes in the hills when he saw two guys pull up on an ATV, bury something that looked like an airplane part, and ride off. Stucky went to the site, removed a piece of scrap metal, and found a missing feather actuator at the bottom of a hole.

But these outings were about more than missing parts; they offered Stucky a kind of therapy. He had lost dozens of friends over the years. He once watched a buddy crash short of the runway at Edwards Air Force Base, killing himself and a crew member. But this was different. Alsbury was more than just a wingman or a squadron mate. Stucky felt like he lost a brother. "Nothing's affected me quite like his death," he said.

Others shared his grief, because they all saw Alsbury as a rare breed. He stood out among the clique of hotshots for his humility and patience and graciousness; he taught several colleagues how to fly. In a condolence letter, June Scobee Rodgers, the widow of *Challenger* commander Dick Scobee, said Alsbury's "courage, intelligence, and good nature" would be "celebrated for as long as we dare to expand the boundaries of our universe."

She added, "From the energy of grief, the phoenix will arise with even more resolve and commitment."

"BAD DAY." WHEN Branson read the message from Whitesides he knew what it meant even before he knew what it meant.

"On my way," Branson replied. He grabbed his passport and hurriedly packed a bag. A speedboat whisked him to the nearest airport. He stopped in Miami for fuel, then jetted on to Mojave.

He read the latest reporting; every major news outlet was on the story. The coverage was often breathless, about how Virgin was building "the world's most expensive roller coaster, the aerospace version of Beluga caviar," and how someone was now dead "in the service of a millionaire boondoggle thrill ride." Joël Glenn Brenner, a former *Washington Post* reporter, went on the air and claimed falsely that the crash was the result of a faulty rocket: "The rocket stopped, and then within one second it restarted, and that is when it exploded." Brenner held Virgin responsible, saying, "They took this pilot's life."

Branson wanted to host a press conference when he landed in Mojave to get out ahead of these falsehoods and retake some control over the narrative; he had his share of detractors but he was generally nimble when it came to managing the media and he didn't like letting others tell his story. Now was not the time, Christine Choi, Branson's longtime adviser, argued. Who cared about public opinion? she said. His employees needed him more. They were a workforce in great distress. A pastor and trained trauma expert was already on his way.

Branson heeded Choi's advice. He met with the pilots and later hosted an all-hands meeting in the hangar. It ended with a group hug.

When he finally gave his press conference he was humble. "We fell short," Branson admitted. But he was not about to quit. He was determined to "learn from this and move forward."

In private Branson was more solemn. Customers were demanding refunds. Sponsors were backing out. He contemplated whether Virgin Galactic could survive. He had previously said that NASA might have lost 3 percent of its astronauts but "a private program can't afford to lose anybody, really." That night he and Whitesides met alone. Branson was ready to find the money to push on. But what if they had another bad day, he asked. Could they survive a second accident?

Neither offered an answer.

STUCKY TOOK IT HARD, but not so hard that quitting ever crossed his mind. An expectation of sudden death came with the job; test pilots learned to metabolize mortality differently than the rest of us.

Alan Shepard, the first American to go to space and one of the Apollo hopefuls, was once asked whether he had second thoughts about the mission after Apollo 1, when three friends got trapped inside the capsule as a wire sparked and caught fire; Shepard could hear them burning alive and screaming "Fire in the cockpit!" No second thoughts, he said. Accidents happened: "That was the way things go in our business."

Stucky's mission was different, though, because his business was a *business*. He read all the hit pieces on Branson and Virgin and Scaled, the ones with headlines like SPACE TOURISM ISN'T WORTH DYING FOR. They could say whatever they wanted about the follies of shuttling wealthy people into space, but he got angry when they implied that Alsbury died for nothing. It showed just how little the authors understood about a test pilot's psychology. Alsbury *wanted* to be in that seat; he had raised his hand and trained and made sure his kids were there so they could see their dad zooming overhead in a rocket ship. How dare someone pretend to know the worthiness of Alsbury's death? Alsbury and Stucky and the other pilots had all signed the waivers, the ones that acknowledged they could die and that their spouses couldn't sue. They knew the risks.

One journalist got under Stucky's skin more than the others. Doug Messier lived in Mojave and spent his days in the airport restaurant, the one with a sign that read AVIATION SPOKEN HERE and an SR-71 pinup on the bathroom wall. A middle-aged frumpy sort with a mop of gray hair, Messier had moved to Mojave in 2011 to write a book about XCOR, another suborbital space tourism company, but the company stopped talking to him, and his book stalled. Now he eavesdropped on pilots eating breakfast, gathering information to write articles for his blog, Parabolic Arc.

Messier had it in for Virgin, or so it seemed. He thought they were dangerously naive—a "hype factory," as he put it. But he had sources, and he was on the ground. He helped a British author write a critical biography of Branson titled *Behind the Mask*. And Messier was the one who interviewed the FedEx driver on the day of the crash.

Normally, Stucky ignored him, jeering his blog as "Parabolic Snark." But after Messier published a number of posts that accused Virgin and Scaled of being reckless and cavalier, posts that the companies left unanswered, Stucky decided to mount his own counterattack. He signed in as "4ger" and replied in the comments section: "In the last couple of weeks I've lost a great friend and a great spaceship. Almost immediately you and other pseudo-journalists were quick to lay blame on the rocket motor and perceived programmatic pressures." Both of those, he said, "couldn't be further from the truth."

He went on, "You act horrified about unconfirmed testing schedules and yet never compare them to other manned rocket programs—either historical or planned." All of them suffered tragedies and setbacks. Rockets were fickle beasts, and the few test pilots crazy enough to fly them knew the risks they were taking.

STUCKY ATTENDED SEVERAL memorial services for Alsbury over the coming weeks. Colleagues gathered for a small ceremony in Mojave and afterward, as a tribute, released dozens of red balloons into the sky.

A larger service took place in the nearby minor-league baseball stadium. Chairs were set up in the infield. Saling and her kids sat behind home plate. A bagpiper wore a bearskin and played "Amazing Grace." Branson attended, quoting Ralph Waldo Emerson: "It is not the length of a life but the depth of life." A flag that flew over the US Capitol on the day of Alsbury's death was folded crisply and presented to Saling in a triangular pine box.

Stucky spoke on behalf of SETP, which Stucky presided over and into which Alsbury had recently been inducted. "We've received countless messages from those in the flight test community that knew Mike," said Stucky. He shared one from the chief test pilot of a military squadron known as the Salty Dogs, who said he considered Alsbury "a Salty Dog and a friend," and who thanked Alsbury for having the courage to "push the limits of the very world we've defined," adding, "There is no mission more remarkable than the pursuit of exploration."

After the eulogies, WhiteKnightTwo appeared in the sky, roaring overhead. In the aftermath of the jet's fading engines, Saling shuddered with tears. They had reached that moment that occurred in every service when the tone shifts from celebrating a life to reckoning with the end of it, the moment that dawns on survivors that life must go on.

Those were long and emotionally wrenching days. When Stucky got home one night he saw a message from Dillon. "Glad you are OK," it read. "I was tripping out for a bit. Sending my condolences for you and your colleagues."

Stucky hadn't heard from Dillon in almost a year. He had almost given up hope of them ever having a relationship again. On Dillon's birthday, Stucky had sent Dillon a note, writing, "In the absence of hearing from you I am just assuming that no contact is what you want for the future."

Seeing Dillon's message now should have lifted Stucky's spirits and given him a reason for optimism. But the crash left him feeling exhausted and cold inside.

"Hmmm, let's see," Stucky replied. "Son ignores father and refuses to include him in his life but would be 'tripping out' if father might be dead. Sure sounds logical to me."

Silence resumed.

14.

THE RED BALLOON

A FEW WEEKS LATER A visitor knocked on Stucky's front door. It was Virgin's then vice president, Mike Moses—an earnest, over-educated man with Apollo-era hair. Moses had grown up in Pennsylvania coal country, but fled when he could to study astrophysics and rocketry at Purdue. He was book smart yet scrappy, hobbling into a seminar one day on crutches from a hockey injury.

Moses left the Midwest and went to work for NASA but held on to a certain folksiness. He wore indigo blue jeans with his cell phone holstered to a braided leather belt; and where others *solved* problems, he *licked* them. "We thought we had it licked, so we're going to take our time to make sure we do have it licked," he once said, after a fuel leak delayed a Space Shuttle launch.

They loved him at NASA. He was a star. Sometimes he complained about the organization's sluggishness, but his colleagues were passionate, and he believed in the mission. On launch days he would drive to work in the middle of the night on the causeways that criss-crossed the marshes around Cape Canaveral. He would often pull over and eat his favorite breakfast, a Mountain Dew with a straw-berry Pop-Tart, while gawking at the site of the Shuttle perched on its launchpad and lit up with xenon. "I can't see how anyone who comes down here to see a launch doesn't choke up and swell with pride," he would say.

But he had endured tragedy and hardships, too. He was leading the propulsion team when *Columbia* crashed, and had spent months after that scouring the forests in east Texas for lost debris. He beat himself up over the accident because he and the team should have known that a gash on the leading edge of the wing would cause a problem during reentry. And he was there when the recession hit and budgets shrank and the Space Shuttle program was formally retired. It was heartbreaking to witness his team's morale slump; these were mission-minded people suddenly robbed of a mission. "The work force down here definitely has me concerned," he said in January 2010.

By 2011 Moses was wondering what he might do next. Then his wife, Beth, heard that Virgin Galactic was trying to hire a director of operations. He and Beth had met in the engineering department at Purdue, where she also studied aerospace. She encouraged Moses to apply, and he got the Virgin job. Leaving NASA was hard. Plenty of his colleagues thought these other billionaire-owned space companies were destined for failure. But Virgin needed someone like him if they stood any chance, and Moses figured he had nothing to lose. "I'm going to take another shot at it here in the commercial sector," he said.

Moses waited for Stucky to answer the door—a rare moment of calm in what had otherwise been the most frantic and exhausting few weeks of his life. *Columbia* was tough to handle, but he had thousands of colleagues, and the US government, to lean on and share the emotional load. Now, because he had gone through all that, his Virgin colleagues looked to him for answers; in Whitesides they saw panic and fear, in Moses they saw resolve. Moses felt like he had no choice. It was up to him to honor the dead but pay equal attention to honoring the living. He was the one who called the pastor in from Houston. He and the pastor had done this before, after *Columbia*. They had shared that emotional foxhole.

And yet Moses was human, too. He felt what others felt, he just chose not to show them, instead projecting strength. He would wait until he got into his car to drive home after another twelve-hour day—that's when he could lock the doors and roll up the windows and cry.

Stucky welcomed him inside.

STUCKY FELT AN adulterous pang letting Moses into his home. He knew, as he showed Moses into the living room, that this was not an innocent visit. For months the two men had been involved in secret discussions about Stucky quitting Scaled and coming to fly for Virgin, about what that would mean, and about how they would break the news to Scaled, because it felt a bit like leaving your wife to marry her sister.

But that sister would give Stucky the chance to fulfill a dream. According to Scaled and Virgin's arrangement, Scaled was to hand over SpaceShipTwo and its test program to Virgin at the end of the year, no matter what, which meant that Virgin would henceforth be in charge of pilot assignments; if Stucky stayed at Scaled he would never fly the ship again. He agonized over the decision but hardly told anyone, except for Agin and Alsbury, because he didn't trust anyone else to keep the secret. The last thing he wanted was for word to leak out, Moses's offer to fall through, and for him to lose his current job, too.

The crash, ironically, made Stucky's decision easy. He looked around and saw a tale of two hangars. In the hangar where Scaled once kept SpaceShipTwo there was now a pile of rubble and wreckage—a stinging reminder of failure. Virgin, meanwhile, had a partially assembled husk of a new spaceship on scaffolding—a glimpse into the future and better days ahead. Scaled had nothing to hand over anymore, and Moses was about to raid one of the most valuable assets it had left: Stucky.

Moses had come to make it official, to offer Stucky the job that they both knew he wanted. In spite of tragedy, Moses was brimming with optimism and a conviction that Virgin was going to emerge from this calamity a better company. He wanted Stucky to help lead its revival. He had a crack corps of test pilots—Dave Mackay, the Royal Air Force veteran; two former US Air Force testers, Mike Masucci and Todd Ericson; and Rick Sturckow, a former marine and astronaut who'd logged four Space Shuttle missions and more than twelve hundred hours in space. Still, nobody knew SpaceShipTwo better than Stucky.

In one way, he was on the cusp of proving his dad wrong: you *could* become an astronaut without the military. Yet his and the other four pilots' backgrounds vindicated his father, because these pilots were civilians now but military men at heart. Space travel had always been like that. Astronomers and entrepreneurs and scientists cloaked their celestial interests and explorations in peacenik language, but there was often a hidden, camouflaged hand pushing them forward. Galileo is often credited with inventing the telescope, but it was a glassmaker from the Dutch city of Middelburg who, a year before Galileo's breakthrough, slid a pair of round lenses into a tube that he lent to a prince on his way to wage war against Spain. It, like other early telescopes, was a tool for spying, what astrophysicist Neil deGrasse Tyson called "terrestrial devices meant to be turned toward the sea rather than the sky, and to be used in the daytime rather than the night."

Stucky had already made up his mind. In for a penny, in for a pound, as he liked to say. He accepted the fact that they would be grounded for a while as the engineers broke out their tools and sketchbooks. It was natural that every good engineer thought he or she could improve upon an old design. There were a few things Stucky thought they should change, but he loved that ship and if there had been another one like it he would have happily climbed in, shaken off his fear, and flown it the day after the crash, just to see what they could learn.

Fly, test, notate, adjust.

That was the credo that all the legendary testers lived by—that oftentimes the best way to break a heedless horse was to ride it again. "Give me a warm day and I'll fly that sonofabitch," Chuck Yeager said about the *Challenger*, shortly after the accident.

So that was that. Stucky was moving on. He felt joy but also some melancholy because what seemed like the end of the world only a month ago, with the death of his best friend and the apparent death of his lifelong dream, was beginning to seem like a momentary pause in the momentum of life. The program was not ending. Another spaceship was being built, and he was going to fly it. The only thing truly gone was Alsbury.

Stucky walked Moses to the door. On the way they passed a deflated red balloon hanging on the wall; a few days after Alsbury's memorial service, when attendees released all those balloons, Stucky and Agin were driving down a desert highway when they saw one snagged on a tumbleweed. They stopped, picked it up, and took it home. Someone had written on the balloon with a magic marker: "Your spirit will always inspire us to new heights."

PART TWO

FATHERS

15.

A RESEMBLANCE

TWO WEEKS AFTER THE crash I pulled into Mojave. I had emailed my editor at the *New Yorker* upon learning of the crash. I proposed writing something about the mysterious pilots testing a billionaire's supersonic, handmade rocket ship out in the high desert of California. It felt retro and zany and endlessly fascinating.

My editor agreed and asked whether we could get real, unscripted access. Despite Branson's aura of spontaneity, his handlers and advisers tended to keep reporters on short leashes. I went to Mojave to see how long that leash could get.

I drove up from Los Angeles. It took about two hours but felt like a time warp: one minute I was in traffic surrounded by juice bars and muscle cars, the next I was barreling down a deserted highway lined with billboards advertising accident attorneys and varicose vein solutions. The road winds through craggy canyons striped with sedimentary rock and across barren plains dotted with semaphoric saguaros. Near Mojave, I saw hundreds of towering wind turbines on the slopes of the Tehachapi Mountains, pinwheeling in the breeze like a field of industrial sunflowers.

I met with Mike Moses and explained how I wanted to embed with his company as it built another spaceship and returned to rocket-powered flights. Like Buzz Bissinger had done with the high school football team for *Friday Night Lights*, I said. I knew this was a big ask. A football coach had fragile egos to protect, while Moses had to worry about corporate secrets and all the risks that came with allowing an outsider access to inside information. Frankly, letting a reporter embed with a high-performance flight-test program was unheard-of. "Keep that man away," Neil Armstrong once said about a journalist. "He's a ghoul."

But Moses had been around long enough to trust his own instincts. "My calling card was built long before I got here," he would say. Until now, most of Virgin's PR pushed glitz and celebrity. But Moses had techs and engineers and pilots with greasy hands and heavy hearts working over-time in the middle of nowhere, trying to do something no one had ever done before, and that was worthy of documentation, he thought. Branson agreed.

Moses was admittedly cautious, wary of promising too much. I told him what I tell other subjects and sources at the beginning of a project—that he didn't owe me anything; that his comfort around me would determine my access; and that while I hoped to become a fly on the wall, that was up to him.

We shook hands, and Moses invited me to come back in two weeks. He would take me around the hangar then, he said, "to show you stuff no one has ever seen."

I PARKED OUTSIDE the main hangar, a huge corrugated metal box they called FAITH—the Final Assembly, Integration, and Test Hangar. It was a clear blue December morning, and the lot was full of trucks and jeeps whose owners drove in early, coming from as far as forty or fifty miles away. Working in Mojave was one thing, but living there was another, with its wretched schools and dilapidated parks and the dollar store parking lot strewn with abandoned shopping carts.

Moses commuted from Palmdale, near where Stucky lived. Palmdale had a shopping mall and some decent restaurants. You could get to L.A. in an hour, on a good day. But it was far from idyllic. A mountain lion attacked and mauled a local auto mechanic, and the city had problems with gangs and hate crimes, earning a reputation as Los Angeles County's "estranged backcountry stepchild." Palmdale appeared on a list of the fifty most miserable cities in the United States. Child abuse hotlines struggled to keep up with calls. The mayor was charged with embezzling half a million dollars from a NASA-funded nonprofit. (He has maintained his innocence.) "No one can get out of here," said one resident. "It's like we're all stuck."

The receptionist took my name and called Moses to inform him that he had a guest. As I waited, employees dropped by the reception desk for small talk and a taste of bite-size chocolates from the bowl that the receptionist kept filled on the counter. Moses appeared, and I followed him down a hallway, through a pair of double doors that required a badge, and past a first-aid station. We entered the bright white hangar. Power tools whirred in the background.

We crossed the hangar, to where a dozen engineers and technicians were gathered near a patchwork-colored fuselage. They faced a whiteboard graphed with dates and part names. A former marine named Bryon Holbrook waved at the whiteboard and then at the scaffolded spaceship. "It's go time," said Holbrook, trying to rally his team. But he saw doubt in their eyes and was wondering how to motivate them.

It was easy for Holbrook. He had deployed with a squadron of warplanes during the conflict in Kosovo, but he told his wife Brandi that building the spaceship was the most fulfilling thing he had ever done in his life. She was happy for him. But he also promised when they moved to Palmdale as newlyweds that they weren't going to stay more than five years, and his time was almost up. Now he was pissed because he felt like she was forcing him out when they were *so* close.

He tried to make his case for staying longer. Think about the Wright Brothers, he said. They crashed plenty, or, as Holbrook put it, "They jacked shit up." At one point Wilbur was so despondent that he declared, "Not in a thousand years would man ever fly." But he and his brother always got up to do it again. If Holbrook left now, he felt like he would be quitting, letting the crash define him. "Are we really going to go to Alsbury's kids and say, 'Your dad died for a program that ended'?" he said. "That's horse shit. No way."

Holbrook walked to another hangar. There, techs used lasers to cut carbon-fiber sheets into shapes for making the ship—wings, tail booms, horizontal stabilizers. They cured the cut pieces in a Celotex oven. Clearing his throat, Holbrook gathered his team around to ask whether they had any questions. You know, he said, "like a rumor you want me to squash? Or solidify? Anything?"

People had been whispering about Branson pulling out. It wasn't an entirely baseless rumor. The board of directors had long been wary of doubling down on space tourism when they could be investing that money elsewhere. Even Branson admitted that, had he known in 2004 what he knew now, "I wouldn't have gone ahead with the project." He explained: "We simply couldn't afford it." But hindsight was twenty-twenty, and so on they went.

A female tech in a Tyvek suit pulled Holbrook aside. She said she needed more people. She was short-staffed and missing deadlines. Missed deadlines meant delays, and those delays added up. When Holbrook told his wife to give him five years, he was certain he could achieve what he came to do in that time. But all those small delays and that one big crash now made it impossible, and Brandi Holbrook was done. She hated hearing the wolf whistles when she dropped their kids off at school. So when Holbrook mentioned staying in the area until Virgin built another ship, Brandi screwed up her face in anger and said, "You keep extending our sentence."

Her frustration grew when a company in Denver offered Holbrook a job after the crash that he turned down because he had come to build a spaceship; he just needed a couple of more years.

But not long after he turned down the offer, Holbrook called me from the road. He was driving a U-Haul across California: the company in Denver had made him another offer. This time, he accepted. "I was going to lose something," he said. "Either the spaceship or my family."

I MET STUCKY on that trip. He recommended the bar at the Broken Bit, a saloon-themed steakhouse near his home. He was waiting at a high-top table when I arrived. He kept an eye on the door because he wasn't supposed to be speaking with any reporters; technically he was still working for Scaled, whose blanket policy was to refer questions to the PR department, to let them decline comment. But he knew he was fixing to leave and that magazine writers didn't come around too often, and he was curious what I had in mind.

Truthfully, it wasn't much. I didn't know the first thing about rockets or flight dynamics or the commercial space industry. Nor did I know much about Stucky. Just that he was a former marine who flew the first three powered flights and that when nobody else at Scaled would talk, he was offering to meet me for a drink. That told me enough, like he either didn't care, or that he trusted himself more than he trusted some suit in the PR department to tell him what he should and shouldn't say.

He ordered a shot of whiskey, neat. He clenched his jaw when he spoke and hunched his shoulders like something was weighing on them, a heavy knapsack or a shell or a lifetime of stories of which he was eager to unburden. I asked questions about the crash, but he was cagey on the details. He could get in real trouble for commenting on the crash before the NTSB completed its investigation, he said, even though he already knew what happened.

He stared hard at me but looked like a man with secrets in search of someone to share them with. He said I reminded him of a fighter pilot he knew thirty years ago but hadn't seen until recently, when Stucky saw the pilot on TV wearing a uniform with three stars on each shoulder—the general in charge of all the airplanes and helicopters in the Marine Corps. Stucky remembered being a young lieutenant when this other pilot was a captain, the best in the squadron, a hotshot who wore aviators and dangled a loopy-ashed cigarette from his lips that stained his mustache. This was the kind of gifted pilot and officer Stucky wanted to become, full of confidence and conviction, one who "never met a rule he didn't mind breaking."

Stucky had a good reason for noting the resemblance. That man was my father.

16.

JOUSTING

MY DAD SIGNED UP TO join the Marines a few months after the fall of Saigon. He was a history major with long hair who drove sports cars and listened to rock and roll. But he felt like he had missed his chance. He played college rugby and was a fierce competitor; he couldn't imagine a greater competition than dogfighting against an enemy airplane.

A magazine article fueled his hunch. The article was by Tom Wolfe and titled "The Truest Sport: Jousting with Sam and Charlie." Wolfe was researching *The Right Stuff* when he wrote a piece about the F-4 pilots dodging surface-to-air missiles, or SAMs, in the fog-socked skies over the Gulf of Tonkin, which Wolfe described as "like trying to fly through a rainstorm without hitting a drop." Here Wolfe first flirted with the idea that among elite aviators, a rarefied few possessed something that eluded the others, that "within the fraternity of men who did this sort of thing day in and day out—within the flying fraternity, that is—mankind appeared to be sheerly divided into those who have it and those who don't."

My dad graduated from Officer Candidate School, The Basic School, flight school, and the preeminent fighter pilot school, Top Gun. He earned his call sign when he was launching off the deck of an aircraft carrier and his engine "rolled back"—suddenly losing thrust. He was about to sink into the ocean when his engine resumed full power, sending a glorious rooster tail over the back of the aircraft. He was henceforth known as "Rooster."

Robert "Rooster" Schmidle.

In 1982 he got orders to be an instructor at the training squadron in Yuma, Arizona, where freshly minted pilots went to learn the ways of aerial combat. My dad didn't fly against the students much, because it wouldn't have been fair. Even when he flew against the other instructors it didn't seem fair. Doug Jones, a radio intercept officer in the squadron, said, "Rooster could do things with the Phantom that shouldn't have been possible, based on the aerodynamics. He could make that thing dance."

Jones flew with a lot of pilots; the F-4 had two seats; the RIO sat in the back. But hardly anyone flew like my dad. Others got twitchy flying low; it was unforgiving down there, the margin for error was tiny. If you glanced at the instruments a second too long you could hit the ground. Another fireball lighting up the predawn sky. Another name across the top of another accident report. But if you were good, it gave you an edge, suggesting that maybe you had *it*, that you could fly at the corners of the envelope and live to tell others about it.

One morning, my dad took off with Jones in the back seat. He stayed treetop low over the main road that ran onto the base. It was early and no one was on the road, or so they thought: my dad engaged the afterburners. Neither he nor Jones saw the motorcyclist, or knew that the man on the motorcycle was the base operations officer. The exhaust wash nearly knocked the ops officer off his bike. He was livid, and when my dad landed, he and Jones got called into the squadron commander's office, where they were chewed out and threatened with having their wings yanked.

That was one of those stories that got funnier with time and alcohol, or when Jones retold it to the wide-eyed students back in the ready room. The students would listen and take notes and wonder if they had the guts to apply the afterburners at that altitude. Some of them were good, a few were great, though many of them faded out after a few years.

One fresh-faced lieutenant was convinced that he had *it*, that big things awaited him.

His name was Mark Stucky.

STUDENTS ROTATED THROUGH the squadron every few months, so Stucky wasn't there long. But my father made an impression because he flew hard—"straight lines and square corners and max g's," as another former student recalled—and didn't care what others thought. He knew what he wanted. Some found his conviction intimidating. My father wasn't one to ask for permission. Stucky said, before he and I first met, "I expect [Scaled] will be as thrilled to see me in front of a reporter as the Commandant would have been to see Rooster in front of one thirty years ago."

While Stucky left Yuma intent on becoming an astronaut, space travel didn't capture my dad's imagination the same way. He was more interested in flying a jet through a canyon, or a barrage of enemy missiles, than floating in the vast emptiness of space. He had a jug-eared friend he met in Yuma named Manley Carter, better known by his call sign "Sonny," who could have done anything he wanted; he had played professional soccer while attending med school. Carter flew in fighter squadrons for a few years before going to test pilot school and becoming an astronaut. Though my dad never understood why someone would want to leave a combat squadron, he and Carter stayed in touch.

In late 1989 Carter launched into space. My parents drove down to Cape Canaveral to watch. We were living in base housing in Beaufort, South Carolina, where fighter jets constantly roared overhead. THE "NOISE" YOU HEAR IS THE SOUND OF FREEDOM, read the sign at the entrance to the base.

A few months after Carter completed his mission, my dad came home from work and told my mom to turn off the news. It was August 1990, and eight divisions of the Iraqi Army had just swept into Kuwait, seizing airports, palaces, and oil fields. George H. W. Bush warned Saddam Hussein to withdraw, or else. My dad told my mom not to worry. She tried. Then she came home and saw his gas mask piled up with his socks and underwear, and heard Hussein threaten to try American pilots as murderers instead of POWs, sending them home in "shrouded coffins." A few days later, my dad and his squadron, VMFA-333, flew their F-18s to Bahrain, in the Persian Gulf.

He was about to get his chance after all.

On January seventeenth my dad led a fifty-plane strike force into Iraq. They flew without lights to avoid detection, but the Iraqi air defenses were waiting for them. It was like flying through a fireworks show as they jinked—"Break right!"—and tossed chaff and flares to decoy the SAMs, while avoiding the antiaircraft fire exploding around them.

My dad kept his eyes on the twinkling lights of the oil refinery in Basra he was assigned to destroy. His bombs hit the target, and he received a Distinguished Flying Cross for the mission, citing his "superb airmanship and undaunted courage" amid "deteriorating weather, unceasing anti-aircraft artillery, and intense concentrations of surface-to-air missiles."

He sent a letter to Carter, who was in France for NASA, a few days later. "You'd rather be here," he wrote. "Remember what 'they' used to tell us about not dueling with SAMs and AAA? Well, today, I bombed a 37mm AAA gun that was almost as accurate as I was. Yesterday I had an I.R. SAM launched at me (flares did the trick)."

He was jousting with SAMs, and it was more thrilling than he ever imagined. "This war is the biggest Rodeo show I've seen yet," he wrote.

Thousands of other marines were in Bahrain, ready to seize their chance. Stucky was a test pilot based in China Lake, California, but happened to be in Bahrain showcasing new flight software in development when the war broke out. He flew a few combat sorties before his commander called and asked him what the hell he thought he was doing, ordering him back to China Lake.

Stucky discovered years later, during a marriage counseling session, that his wife, Joan, had been the one who alerted his commander. She insisted that she did it for all the right reasons, that if he died in combat he would never fulfill his dream of becoming an astronaut. And wasn't that why he joined in the first place? Still, Stucky was upset that she had the gumption to do something like that. In his mind there were no guarantees in life, and who knew what might happen tomorrow.

The war ended in February. Two months later my dad flew home. My mom and my brother and I put on our Sunday best and gathered on the runway holding miniature flags as the band played, and when the jets appeared in the distance we waved the flags, and when they landed we ran to my dad.

The next day Carter boarded a twin-prop commercial plane in Atlanta on his way to Brunswick, where he was supposed to give a speech. People loved to hear astronauts talk about the hurdles they had overcome and what it felt like to float around in orbit and look down on our Pale Blue Dot. Carter was scheduled to return to space in a few months, and who knew what he had coming after that—maybe a run in politics, with his easy smile and previous turns as a fighter pilot, doctor, test pilot, professional soccer player, and astronaut.

But on the approach to Brunswick the plane rolled hard and suddenly to the left. It nose-dived into the ground. Everyone on board was killed, a cruel and random end to a life lived at the edge of the envelope.

My father flew to Houston for the memorial service. Congregants sang a version of an old British naval hymn, "Eternal Father, Strong to Save," in which they prayed for protection and mercy and inspiration:

O hear us when we cry to Thee,
For those in peril on the sea!
. . . Be thou shield forevermore,
From every peril to the Corps . . .
O hear us when we seek thy grace,
For those who soar through outer space.

17.

WHALE LINES

HALFWAY THROUGH *MOBY-DICK*, after the crew mistakes a giant squid for the great white whale, Ishmael slips into a meditation on the rope that ties the harpoon to the ship.

A whale line, he says, is a "magical, sometimes horrible" thing—extraordinarily strong for its slim girth, and the only way for a whaler to reel in his catch. But if you lost track of its "hempen intricacies" as the harpoon flew and the rope snapped taut, its "horrible contortions put in play like ringed lightnings," then that whale line was more fatal than gunpowder.

"All men live enveloped in whale lines," Ishmael says.

I read this to mean many things.

That it's not the sharp end of the harpoon that threatens us most, but the innocent ropes by our feet that present the "silent, subtle, ever-present perils of life"—the feather designed to save SpaceShipTwo from certain peril that becomes the cause of its demise; the commercial plane that crashes and kills an astronaut.

And that we are products of our past, entangled with kin by "complicated coils," stronger than we care to admit.

WHEN I FIRST met Stucky he felt instantly familiar, like I knew him from another life, though maybe I just knew his type.

He would call me on his commute to work, and we would talk for an hour until I heard him turn off the engine. I knew he was sitting in the parking lot, waiting to go inside, often explaining something he didn't have to explain, like why he refused to give up his paraglider even after he broke his back and Joan threatened to leave him.

Stucky didn't have to explain because after my dad crashed his motorcycle for the umpteenth time, he swore to my mom that he would quit and hang up his leathers, but then his bones healed and his ego recovered and he was out there again, twisting the throttle as he came out of a turn. He did so not as an act of defiance against my mom but just because that's who he was. Mediocrity bored him.

I rebelled against my dad, like many other sons. He would deploy for months and was often emotionally distant when he was home; he disciplined through disinterest and disappointment. He had firm opinions about manhood. He often rolled his eyes, like he had never been a teenager, so I tried to remind him what it was like: I hung out with the wrong crowd, did a bunch of drugs, and skipped school. On a field trip I pierced my ear with a diabetes lancet; my dad ignored me until I removed the stud a week later. I told myself that I was pushing the envelope, but I was really just pushing for his attention. It went on like this for years; I crashed his Porsche into a ditch and was arrested a couple of times. His friends used to say that I reminded them of him, and I fooled myself into believing that I inherited something special, when all I got was his ego without any of his conviction.

I went into journalism after college. I wanted my own adventure. The combination of plotting terrorists and young Americans dying in foreign wars had created an appetite for international news, but there were only so many foreign bureaus. I saw freelancers breaking into the business by taking extraordinary risks and wondered if I could do the same.

I went to Iran, studied Farsi, and wrote a few articles when I got back.

I wandered around Central Asia, took notes, and wrote a half dozen more.

After that I moved to Pakistan for two years. Even there, thinking back, I rebelled against my dad by trying to get inside the Pakistani Taliban—as a reporter, of course, but also as someone who wanted to understand the *other side*, these young men with Kalashnikovs who snuck into Afghanistan on goat paths in the middle of the night to attack American soldiers and marines, like my dad and my brother and my cousin.

My dad and I argued about who knew more, me or the CIA. I thought I did because I had just come back from the mountains along the Afghan border, where, to blend in, I would color my hair brown and speak Urdu and wear local clothes, including those strappy sandals with a heel, and where the police had abandoned their posts, and where I met guys who carried rocket launchers and claimed that they knew Ayman al-Zawahiri.

Years later, after I'd moved back home and started writing for the *New Yorker*, my insecurities remained. I was writing about soldiers and SEALs for the magazine, but when asked whether I was covering the military beat, I said no, that "I don't want to be the military guy," because I wanted to be my own guy. I failed to appreciate the sturdiness of those whale lines.

I knew there would come a time when I would be ready to write about my dad, about growing up in the shadow of this extraordinary man who flew fighter jets, and raced motorcycles, and hunted for wild boars with a longbow and home-fletched arrows, and earned a PhD in philosophy, and wrote his thesis on Wittgenstein, and lectured at the Sorbonne, and retired as a three-star general, which he took hard because he wanted four.

But he was still alive, and it seemed impossibly difficult to write about someone so close who was still alive. Legacies were fluid, funny things. I didn't see how to do it, how to hit pause on life in order to study a few key frames, to freeze a relationship long enough to make sense of it. Like Kierkegaard said, you could only understand life by looking backward but had to live it looking forward.

Then I started working on this story.

I would call Stucky's friends, some of whom would stop me when they heard my last name, and ask me whether I was Rooster's son, and then insist on telling me one story or another. My dad wasn't the type who told stories. I recently found a letter Barack Obama wrote to him among a stack of papers in my parents' basement; my dad had never mentioned it.

But one story did come up now and then. It was the one that confirmed my father's legend. And while I had heard snippets over the years—something about a bridge in Bosnia and him blowing up a tank carrying some genocidal maniacs—I didn't know the details.

Only now was he ready to dig out the cockpit video and tell me what really happened.

ON THE MORNING of April 4, 1994, my dad took off from an air base at the foot of the Dolomites. He was dreading the flight. It was his last sortie. He had been sleeping in a tent for three months, enforcing the no-fly zone over Bosnia. But in a few days his F-18 squadron was fixing to go home without having fired a shot, and not for lack of targets; Serbs were slaughtering Muslims in the foggy villages below.

My dad and his wingman flew southeast, over the town of Goražde. Serb tanks and armored personnel carriers had swept into the town from the south, shelling Muslims living north of the bridge. British peacekeepers, holed up in a bunker on the north side, worried that they would soon be overrun.

The UN commander warned Serb general Ratko Mladić to cease shelling or else. Mladić ignored him.

The clouds hung thick and low, swallowing the nearby hilltops. They offered the Serbs cover because what sane pilot would think to fly below the clouds, where one second they would be blind in the clouds and the next they could be a pile of fiery rubble.

They were circling above Goražde when the radio crackled to life: NATO command, asking if anyone cared to swoop down and take a look. Other fliers declined and said it was too low. My dad volunteered. He looked down over his wing at what looked like a woolen quilt below. The airplane did not have GPS, but he'd studied the maps and knew the general layout, how the Drina River there ran west to east, bisecting the town. If he could find the river, he could navigate the rest of the way.

Then he saw a break in the clouds, like a loose stitch, just big enough for him to glimpse the Drina below.

Follow me, he told his wingman.

He pushed the stick forward and dove through the clouds, into a world of zealotry and bigotry and ethnic strife, where the Serbs were raping and massacring and burning babies. When they cleared the clouds he was even lower than he had expected, just a few hundred feet above the river. They skimmed the water, roaring through the snaking, narrow ravine— barely wide enough for him and his wingman to fit without scraping their wings on the canyon walls. One mistake and they would be dead.

Blazing into Goražde, he watched out for SAMs and popped flares to throw off any incoming fire. He saw muzzle flashes on the hillsides. Over the radio, a Brit in the bunker asked my dad if he could see the Serb tank on the south side of the river; the Brit—a forward air controller, or FAC— tried to steer him to the tank, but it was hard to pinpoint because the tank was tucked in a neighborhood.

"I can't see that son of a bitch," my dad said.

"You may see gun smoke when they fire," said the FAC.

My dad was now west of the town and turning back, a bull about to make another charge; at Top Gun, they advised against this because the enemy knew you were coming and it was dangerous to fly over the same target twice. But my dad heard desperation in the FAC's voice ("These tanks *have* to be destroyed") and knew that this was his chance.

On the next pass my dad asked if he was "cleared hot"—authorized to fire—and the FAC said, "Anything to the southeast of that bridge is fair game." He swept low over the spot where the FAC said the tank was hiding and dropped a bomb, but it dudded because he was too low for the munition to properly detonate.

My dad's wingman had seen enough. He reported that his fuel was low and that he couldn't breathe. My dad told him to go up and get above the clouds. He couldn't understand how you could come this far and want to retreat because, as he said, "That's not the way I appear to be hardwired." But others couldn't understand how he stayed down there all alone. People would later speculate whether he had been reckless and foolhardy.

He prepared to make another pass.

He was running low on fuel, and luck, when he thumbed the weapons switch on the stick, activating the Gatling gun in the nose of the jet. If he could get low enough, the gun would chew up the tank, but he would have to be practically on top of it, and then be careful to avoid catching any ricochet. He had only enough fuel for one more run.

This time as he dove he kept his eyes locked on the tank and his finger on the trigger.

Eight hundred feet. Seven hundred feet.

The altimeter spun and he dove and the houses got bigger and he saw the tank on the side of the road with its turret pointed north.

Six hundred. Five hundred.

He pulled the trigger.

Four hundred. Three hundred.

He unleashed all his ammunition. The barrage of cannon fire lit up the tank. *Boom!* It flashed on his infrared display.

Two hundred.

He jerked the stick back in his lap and rolled right to dodge the ricochet.

"Got it!" he said, through gritted teeth, climbing into the clouds and straining as the g's pressed him in his seat.

When he radioed back to the base that his jet needed to be rearmed, the maintenance crew cheered and hollered because this was what they came to do, to stop a genocide, which was noble and yet probably too late. My dad felt good but unsatisfied because the Serbs were bound to come back with more tanks. Radovan Karadžić, the Bosnian Serb president, said, "I am convinced there will be escalation," and a week later, as Karadžić's men got back to their business of bombarding the town, a British jet swooped into Goražde to destroy a Serb tank but was shot out of the sky by a SAM.

My dad flew home the next day and pretended to be happy when he arrived. But it was all an act. He felt the war was about to really kick off, and although he got his chance to prove that he had *it*, he now had to sit home while others got their chance. He felt distant, away from the action, and that made him resentful.

I didn't know any of this at the time. I was fifteen and adversarial and just thought he was grumpy.

TWENTY-ONE YEARS LATER, Dillon Stucky was a struggling actor waiting tables at an Italian restaurant in L.A. when he decided that enough was enough. It was May 2015. He hadn't seen his dad in months, and their last exchange, soon after the crash, ended with Stucky's short-tempered mockery of his son for saying he was tripping out at the thought that his dad might have died.

Dillon sent Stucky a friend request over Facebook.

"Why the friend request?" Stucky replied.

"I am pretty tired of being estranged and want to be able to talk to you and see you," said Dillon. This was not a sudden revelation, he added: "I've been wrestling with the idea of establishing an open line of communication and relationship with you for a very long time."

A part of Stucky felt touched and relieved and ready to do anything to spend time with his son, to take whatever relationship Dillon was offering, because it was better than what they had. But another part, like a whispering devil on Stucky's shoulder, seethed and grew defensive and rehashed the memories of Dillon avoiding him in the courtroom during the divorce hearings and dropping out of the Air Force Academy and ignoring Stucky when he drove hours to watch Dillon compete at college track meets. That part of him wanted to prove something.

"I don't understand why our relationship requires 'wrestling with an idea,'" said Stucky. What was this idea that Dillon found so confounding, he wondered. "If it's that hard of a decision, then I'm either not worthy of your love or else you need help."

Stucky was missing the point. Again. And even though Stucky was no more a typical marine than my dad was a typical marine, it proved that even the ones who fashioned themselves as eccentric exceptions among their leatherneck community were often so callous and driven and intent on proving a point that they forgot what it was they were trying to prove. "You are Marine kids and can chew nails while other kids are sucking on cotton candy," says Bull Meechum, an F-4 pilot and the main character in Pat Conroy's *The Great Santini*. But who preferred the taste of rusty iron over spun sugar?

"This is me ending the 'struggle,'" said Dillon.

Stucky accepted Dillon's friend request, even though it felt like a concession to do so on Dillon's terms, to pretend like those painful years never happened. But Stucky didn't have a choice. He wanted to see his son again. To know him again, to learn what kind of man he had become and what he wanted to do with his life and what he feared and what he hoped for and all that he could learn beyond the scraps he currently saw on social media.

He sent Dillon his phone number but warned him that his phone went into Do Not Disturb mode every night at half past eight. He was busy, in other words.

Dillon finally drove up from L.A. for brunch. It was fun and easy and felt like the old days, and that was part of what bothered Stucky. "How can bloodlines be so easily cut and so easily sutured?" he asked Dillon. "Is our foundation really so tenuous?" He still wanted to dredge up the past. But Dillon knew it was a trap, so he stayed focused on the present, and they talked mostly about the one thing that felt safe: flying.

Dillon wanted to learn how to fly.

Stucky promised to teach him everything he knew.

18.

BLANK PAGES

MIKE MOSES SAW THE CRASH as a chance to start again.

When, in late 2014, he could think past the loss of life and the trauma and the financial costs, and all those deadlines and milestones marked on the whiteboard that he might as well erase, or at least add a couple of years, he tried to appreciate the opportunity. This was a chance to build a better spaceship, to fix all the things they didn't like about the old one.

Moses told his engineers to reassess everything, from the rocket motor and the landing gear to the feather system and the seat belts. Nothing was sacred. It felt liberating to sketch on blank pages.

Everyone respected Scaled and knew its designers were brilliant. But Scaled was a prototype shop, its engineers nourished on the belief that perfection was the enemy of good—an adequate philosophy for building one-offs; a fraught one for producing rocket ships to safely shuttle paying passengers to and from space.

"They were asking, 'Is it safe for twenty flights?' We're asking, 'Is it safe for a thousand flights?'" said Moses.

That was the greatest challenge for him and his team—to make a spaceship supple enough to break the sound barrier, light enough to reach space, strong enough to avoid breaking up on reentry, and rugged enough to make the journey once a week for years to come. Reusing rockets was

Mike Moses.

a relatively new idea. For fifty years, NASA and the Defense Department and its contracted launch providers had chucked used ones into the sea. The Space Shuttle sort of addressed the problem, though it still discarded boosters on the way up, and even the workhorse of the Shuttle fleet, *Discovery*, only flew thirty-nine times. Moses was trying to build a ship that could make at least five hundred trips.

Now his engineers whipped out their notebooks, squiggling and sketching and modeling to ensure that every angle and plane was just as they wanted it to be. This was a benefit of carbon fiber: it was light but incredibly strong—a "miracle material" for aerospace engineers; and unlike a slab of aluminum, it could be sliced and molded to suit a designer's needs. "You can literally tape and glue parts on and off," said Moses.

Scaled had done just that. They were improvisers: instead of building a new wing to achieve some desired dimension, they would slather Bondo onto the wing spar. Slapdash was fine if Stucky was conducting a test and knew what he was getting into, but not if Lady Gaga was in the back.

Some Virgin folks bad-mouthed Scaled for being sloppy. In some cases, they were justified: When Scaled finished testing WhiteKnightTwo and transferred the mothership to Virgin, techs from Virgin had crawled into the wing bays to inspect the airplane and found scores of hairline cracks in the sealant. Virgin's more skeptical engineers said they shouldn't trust anything from Scaled and disparaged Scaled's inspection regime for being random and shoddy, while crowing, "Ours is based on analysis."

Moses was more cautious. He reminded the naysayers that Scaled didn't typically inspect their airplanes because their clients didn't care about longevity; people hired Scaled to prove a concept, and they had proven dozens.

Now Virgin was the one with something to prove.

And for all of Richard Branson's success, he didn't know anything about building rockets. His gift was knowing what people liked. He was a tastemaker. He found musicians and recorded their songs and then sold their records. He spruced up airplanes, trains, hotels, and gyms. But he didn't *make* stuff. He was a marketing genius, not a nuts-and-bolts guy. Now, suddenly, because of an awful accident, he had been thrown into the business of manufacturing spaceships.

He was reliant on people like Moses for success.

One evening, I met Moses and his wife, Beth, at a sports bar in Palmdale. She is blond, with shampoo-ad hair, and several inches taller than him. A NASA veteran herself, Beth had led a team responsible for building the hardware used by astronauts during spacewalks. She was not, in other words, the sort to stare at the sky and fret while others performed death-defying acts; she would have been happy performing the death-defying acts herself.

Designing space hardware was a complex assignment. "You have to understand how hardware behaves when it goes through hot and cold cycles of day and night—and is in a vacuum," said Beth. Power cables often got stiff. Sometimes they shrank. And fluid lines? "The worst," she said. "It would be like if your garden hose was your worst-nightmare high-pressure garden hose, and you couldn't get it to point where you wanted it to. And you're in a spacesuit." To design that, she said, "You literally put a human in a spacesuit in a thermal vacuum chamber and you make the hardware hot and cold and pressurized, just like it would be in space, to learn how it should be built."

She and Moses formed a potent intellectual duo, needling each other like only married aerospace engineers could. "I used to say that his program was just the truck that got me there," said Beth. "He used to say that, without his truck, all my spacewalk hardware was just very expensive scuba gear." She hosted Bitcoin hardware in their garage.

We shared a roomy booth. He drank craft beer. She sipped a fluorescent cocktail from a martini glass.

Beth was used to being the black sheep in a male-dominated world. Virgin's pilots, engineers, and techs were overwhelmingly male. Historically, women have accounted for 20 percent of all engineering majors. "I was sometimes the only woman in a classroom of ten or fifteen males," Beth recalled. After a while? "You don't notice it and you don't care."

Beth acted like one of the guys, though it was hard to say whether her self-deprecations were genuine or a learned defense mechanism. I was with her once when we heard the faint sound of Adele. "Anybody else hear that?" Beth asked, before realizing that it was her cell phone in her back pocket. "I butt-dialed Adele," she joked. "I was telling blond jokes earlier, but it's a long blond joke, so I can't remember it."

At the bar, she let Moses do most of the talking. He was reluctant to discuss schedules and projections but thought it would take at least two years for the program to return to where it was on the morning of the crash. Who knew where the commercial space industry would be by then. There were almost twenty rocket companies in Mojave alone. Most were small companies. Many would likely fail.

Moses hated to see people fail, and there had been some spectacular recent failures. Three days before the SpaceShipTwo crash, Orbital ATK, a commercial rocket company under contract with NASA, was attempting to send its Antares rocket on a resupply mission to the International Space Station when the Antares burst into flames. Even SpaceX continued to face setbacks. In August 2014, one of their rockets exploded above the company's facility near Waco, Texas. And in June 2015, two weeks before I met Moses and Beth, SpaceX was making its own ISS resupply attempt for NASA, launching its Falcon 9 rocket from Cape Canaveral, when the upper stage overpressurized and blew up. All three missions were uncrewed.

Moses ached for them. But he also admitted feeling some vindication, because after each accident the news coverage would invariably feature someone saying, "Space is hard"—a message to naysayers and a reminder to themselves and their propulsion-minded peers. There was a sort of dark comfort buried in failure. "Rockets are a fundamentally difficult thing," Musk, of SpaceX, said after the June accident.

But Moses was equally, if not more, affected by others' success. SpaceX, by that point, had already put a rocket into orbit, deployed satellites into orbit, docked a capsule at the ISS, and nearly landed its own rocket, using a reverse propulsion system and retractable legs, on a platform floating in the Atlantic Ocean. And Blue Origin, Virgin's primary competitor, had recently completed a successful test launch of its suborbital tourism rocket. Moses cursed when he saw the footage; Whitesides, the CEO, tried to downplay the rivalry between Virgin, Blue Origin, and SpaceX, to say it wasn't a new space race, but Moses knew there would be winners and losers. That was sufficient to call it a race. He said of Blue Origin, "I look at their timeline and see that they have a good shot at beating us."

Moses believed that Virgin was offering a more compelling experience— WhiteKnightTwo's traditional takeoff and slow climb; SpaceShipTwo's release, boost, and J-shaped gamma turn into space; followed by that long, corkscrewing descent onto the runway. Moses was biased, but it sounded better to him than sitting in Blue Origin's automated rocket that was up and down in eleven minutes: all action, no foreplay.

"I'd buy a ticket with Blue," said Beth.

Moses turned and looked at her, shocked and disapproving. "You can't say that to a *reporter,*" he said.

ONE DAY, STUCKY took a seat at the end of a long table in the conference room, across the lobby from the receptionist with the candy bowl. He waited for the other pilots to arrive with their thermoses full of coffee and notebooks opened to blank pages.

The last one in closed the door.

It was December 2015. Stucky had been at Virgin for about a year.

The transition had been surprisingly smooth. He didn't miss the ruthless-ness at Scaled, where the other test pilots were mostly young guys eager to prove themselves and who begrudged Stucky for standing in their way. He fit in better here, one of a handful of retired military guys, all of whom had proven themselves in their former lives. "We are just a bunch of old, bald, fat-men trying to finish one last big job before hanging up our flight suits for good," as one pilot put it. They were in this together, and not to stab one another in the back. At least not yet.

The pilots kept a standing weekly meeting with an engineer named Mark Bassett. Bassett had knobby elbows and slide-rule bangs. Everyone called him "McLovin," after the character from *Superbad*, a nickname Stucky coined when he and Bassett worked together at Scaled. Bassett didn't mind it, but even if he did, objecting would have been futile because it came from a pilot, and the pilots were gods. "They walk on water and carry a bus under each arm. They're the ones strapping their ass in the seat and taking the risk. They have to believe they're indestructible," said Pat Marshall, Virgin's director of flight test.

Besides, Stucky didn't bother with people he didn't take seriously, and he thought the world of Bassett; he called him McLovin out of respect.

Respect. That was one thing that upset Stucky. The lack of respect he heard from some of Virgin's upstart engineers, who were prone to sling mud at Scaled about how Scaled cut corners and made an unnecessar-ily dangerous spaceship because they didn't think to design a device to inhibit the feather from unlocking, when, in fact, they *had* thought of it, but they didn't think any of their pilots would be foolish enough to unlock early. "Scaled did things the Scaled way, and Virgin contracted with them to do it that way, but then bitched the whole time about price and sched-ule," said Stucky.

Initially Stucky held his tongue because maybe they were right and Scaled was wrong. Maybe the design needed a thorough reassessment, and bigger horizontal stabilizers, and more safety latches and inhibitors to pre-vent people from hurting themselves. One of Bassett's engineers designed a tiny solenoid pin to keep the feather locked when the rocket motor was firing, and that was fine with Stucky, as it weighed practically nothing and didn't hinder performance. It also satisfied the people wringing their hands about lessons learned and not repeating past mistakes, even if it seemed like they were fighting the last battle rather than preparing for the next one.

But when Virgin finished the new, redesigned horizontal stabilizer, or h-stab, the electric controller failed repeatedly during a stress test; Stucky suggested that maybe Scaled wasn't so stupid after all.

Stucky certainly wanted to fly a safer spaceship. But not at the cost of flying. That's why he and the other pilots toted full thermoses into long, tedious meetings, so they would be ready to go when the time was right. But when? He knew that grounded pilots were grumpy pilots and that a time would come like it came to Ishmael, the merchant sailor, when he had been away from the sea too long and felt tempted to swipe randomly and cruelly at pedestrians, to knock the hats off their heads—just because.

Stucky knew that, if Virgin allowed, the engineers could spend another decade tinkering and spinning over whether some part or another was durable enough or light enough. "There's always somebody who's not sure about something," he said, and nobody wanted to live with another fatal accident on their conscience. But at some point someone would have to draw a line, and they would have to live with what they had, and then it would be up to the pilots who carried a bus under each arm.

That's when these weekly meetings would pay off. They were nomi-nally a chance for the pilots to talk about the layout of the cockpit and whether they should install a toggle switch or a mushroom switch or a pickle switch. But they were really rehearsals, a chance for them to see who laughed at what jokes or how they responded to their own janky PowerPoint presentations. At least for Stucky, for whom one revelation from the crash was the risk of communication breakdown in the cockpit, how muddy language contributed to a failure of what aviators called "crew resource management," or CRM.

Stucky held Alsbury responsible above all for losing focus and unlocking the feather that day. But he wondered what Siebold was think-ing when Alsbury said, "Unlocking," and how come Siebold didn't hear Alsbury and try to do something about it? Stucky hoped, had he been the pilot, that he would have heard Alsbury and had the wherewithal to slap Alsbury's hand away before he could unlock the feather. "I will always believe that if Mark had been in that spaceship, they wouldn't have crashed," said Agin.

So when they sat here, now, talking among themselves, Stucky would note who came prepared or got easily distracted, or worried about the things they shouldn't worry about, or didn't worry about the things they should. They played out scenarios, like what if they were descending in the spaceship on their landing approach and a yellow caution light flashed because they were outside the "glide cone"—an indication that if they didn't alter their approach they could miss the runway.

Like SpaceShipOne, the Space Shuttle, the X-15, and other gliders, SpaceShipTwo descended in a coil shape, using drag to control its speed. Each of those aircraft had airspeed and altitude indicators—"glide cones"— to ensure a safe landing.

"If it's only yellow?" said Mike Masucci, the former U-2 pilot. Not all caution lights were the same, and he felt a yellow glide cone caution was more a suggestion than an order. Red was another story. "Red should scare the crap out of you."

Rick Sturckow, the former marine and NASA astronaut, disagreed. A yellow light was more than a gentle nudge. "It should scare the shit out of you," he said. "It's, 'Holy shit, if I don't do something pretty soon, I'm in a bad place,' because when it turns red it's gonna be too late."

Sturckow was right, said Stucky. And though Masucci's point about the caution light was fair, Stucky worried that if they told themselves it was safe to ignore a glide cone caution, then their brains might ignore another, less forgiving, caution. "You start going down a slippery slope," he said, and he didn't think he needed to remind them how accidents happened.

IN SEPTEMBER 2015, Stucky and Dave Mackay, the former Royal Air Force pilot, drove to L.A. Stucky had suggested that he and Mackay deliver a presentation on the crash at the annual SETP symposium. Pairing with Mackay was his shrewd idea: Stucky had the rapscallion reputation, whereas Mackay, a few years older and more reserved, was a graybeard with a Scottish burr. Stucky drafted their remarks and gave the script to Mackay.

They took the stage. Stucky spoke about how he had imposed "sterile cockpit rules" on the first three SpaceShipTwo powered flights because he "did not want extraneous calls during boost and other high-workload portions"—a veiled jab at Siebold for shouting "Yeehaw!" during the boost on the fourth flight. This, as far as Stucky was concerned, was as unsterile and undisciplined as you could get.

Then Mackay took his turn and explained how, in spite of Stucky's rules on the first three flights, Scaled "gave their pilots-in-command the leeway and authority to run their cockpit as they saw fit," which Siebold did. Siebold's failure to "impose a rigid set of standard operating procedures" created "an environment where callouts and actions could be missed both in the cockpit and in the control room." Which was what happened: Alsbury announced what he was doing—"Unlocking"—but everyone missed it until it was too late.

The stage lights were blinding, and Stucky couldn't see beyond the first row, but he knew Siebold, who had recovered from his injuries, was out there somewhere in the banquet hall. Stucky ran into him later by the coffee urns. Siebold looked furious.

"It's too early," Siebold grumbled, before walking away.

19.

STARMAN

STUCKY PREPARED TO EMBARK on a rescue mission. In February 2016, Kyle Hunt, a young Virgin structures engineer, left Mojave and drove an hour to a trailhead to spend a long weekend hiking alone in the woods.

But on Monday Hunt didn't show up for work. Or answer his phone. Colleagues got worried.

Stucky didn't know Hunt or whether he'd ever had a conversation with him. That was beside the point. Hunt was a colleague, and he was missing. They had to find him.

Stucky organized a team and was about to set out when Steven Avila, a Virgin lawyer, heard of the plan. Avila called Stucky and others into his office.

He asked Stucky what he thought he was doing.

What does it look like? Stucky replied.

Avila encouraged them to stand down; the authorities were on it, he said, and their amateur investigation could get in the way.

This set Stucky off. He was fed up with Virgin's lawyers. They had insisted upon pursuing legal action against Scaled after the accident, citing Scaled's breach of contract, because Scaled was supposed to deliver a spaceship that was now a pile of wreckage in a hangar. Stucky had warned against taking a hard line with Scaled, since Virgin was about to build its own ship and arguably needed Scaled's expertise now more than ever. But the lawyers pressed on and secured a settlement, which might have counted as a win, had it not poisoned the relationship between the two companies.

No way was Stucky about to listen to them now.

Stucky charged out of the room. He did things his own way, but for all of his lone-ranger mystique, Stucky was all about the mission and supporting everyone who was also ready to bleed and sweat for it. Stucky was there, with a bus under each arm, for team outings at the go-cart track and the ropes course, and when heavy rainfall triggered mudslides that turned the roads around Mojave into a silty soup that washed away pets and buried Stucky's neighbor alive in his SUV, it was Stucky who shuttled his colleagues home from Mojave in a small prop plane.

Stucky and two other engineers drove to the trailhead. When they saw Hunt's car, they headed into the woods. Stucky felt déjà vu and maybe that's why he had been so spitting mad. He had, thirteen years earlier, led another search party into the wilderness for a missing friend whom the authorities couldn't find.

In June 2003, Ron Rosepink, an Air Force test pilot, had gone into the woods near Mojave with his paraglider. He never came out. Rescue teams searched the area. Conspiracy theorists speculated that Rosepink had staged his death and fled to Costa Rica or accidentally landed in a marijuana field and was killed. The authorities were concentrating their search as far as fifty miles from where Rosepink had parked his car, on the assumption that Rosepink had drifted in the wind and crashed. Stucky thought this ill-advised and suggested they look closer, that Rosepink was an able pilot, and that it was more likely he had slipped on a mountainside. Stucky felt like he was making headway with the investigation when a ranking officer on base heard about Stucky's involvement and ordered him to stand down. Stucky obliged, angrily.

Months went by. Rosepink's wife, Deborah, was struggling financially because the Air Force took Rosepink off payroll, and his life insurance policies refused to pay in full without a confirmation of death; Rosepink's daughter dropped out of college to help pay the bills. (An SETP scholarship fund enabled her to later resume her studies.) They needed Rosepink's body to move on.

Two years later, in the woods near Mojave, a hiker stumbled upon a human jawbone, and a nylon parachute was also soon discovered: both belonged to Rosepink. Gaining access to Rosepink's file, Stucky helped authorities piece together what had happened, how Rosepink had probably tripped and gone sliding down the mountain on his parachute, like he was riding a nylon toboggan, before slamming into a tree. Stucky had been right all along: Rosepink hadn't drifted or staged his death. Deborah didn't get her husband, but at least she got some answers.

Stucky was replaying all of this in mind as he and the two engineers set out in search of Hunt. He wasn't about to repeat the same mistake. Standing down had potentially cost Rosepink's family two years of added distress, and for all he knew Hunt was dead, too. But if they could find his body it might relieve Hunt's family of unnecessary trauma.

Hiking out, Stucky hollered and peered into the bottom of the deep ravines where Hunt might have fallen. At one point Stucky looked up and saw a rescue helicopter hovering overhead. Officials hung over the skids, pointing down, into a ravine, where they found Hunt's body later that day.

Hunt's parents flew out from their home in the alfalfa country of Kansas, not far from Stucky's hometown. Virgin held a memorial service for Hunt in FAITH, and afterward Hunt's parents drove home with their son's ashes in an urn wrapped in the fleece blanket his grandmother had made for Hunt when he was a child.

"He was living his dream," said his mother.

ON FEBRUARY NINETEENTH, a week after Hunt disappeared, Mike Moses slipped behind a thick black curtain and drew a deep breath. He was forever getting pulled this way and that, trying to be everything for everybody.

He would jump from a phone call with the FAA, to a meeting about emergency protocols, to a discussion about where his assistant should buy doughnuts and breakfast burritos for guests on future flight days, to media training. For instance, did people know how to acknowledge a stupid question without validating its stupidity? "The biggest thing is how to learn how to say, 'No comment,' because there's a 'No comment,' which is, 'Okay, you're not talking about it,' and then there's the one that's, 'Okay, you're clearly hiding something,'" he said.

Moses seemed to have his hands on every lever. He was Lee Iacocca but with the temperament of Ned Flanders, all howdies and smiles, somehow even remembering to give his assistant a card on her birthday. One day there was a potluck lunch, and people would have understood if Moses had just gone to the store and bought something. But he went home and dug through his recipes until he found the one for Pittsburgh-style barbecue that he made in the Crock-Pot with chipped ham, molasses, ketchup, and honey.

He set the tone and urged his team to keep it simple, to speak using real words that meant something and not resort to empty corporate jargon. "I don't want to create acronyms for the heck of it," he said.

But simplicity could feel elusive if you worked for Richard Branson. He, too, was everywhere, trying to be everything for everybody. But what Branson said so often didn't mesh with reality. "One of the things I hate is the world judging us based on what our marketing has said in terms of our readiness to fly and the depth of our knowledge," Moses said to me once. He thought it was unfair to read Virgin's press materials and "assume that's as smart as we are," because all the empty promises and highfalutin press statements made them seem like a bunch of clowns, like they were never going to lick their problems and achieve what they set out to do.

Which was why now, as the sound crew checked the levels, and a fog machine spewed clouds of mist, and Branson was expected to show up any minute with hundreds of guests to stage the kind of spectacle only Branson could stage, Moses slipped behind that black curtain to gather himself for a moment, to remind himself that after everyone showed up and gawked they would eventually leave, and then he could get back to doing his job. On one side of the curtain were rows of chairs and a stage, but on his side it was only him and the spaceship. When he gazed at its shiny coat of paint, he realized that the last time he saw a spaceship in the hangar in full regalia was the morning of the crash, when Alsbury was still alive and the world was a very different place. Seeing it now was like a punch in the gut or, as he put it, "an emotional kick in the shorts."

When Branson arrived he asked for a cup of tea. He was wearing blue jeans, a white Oxford, and a black leather jacket. He traveled light for a man of his means. Perhaps he had taken a lesson from the last extravaganza, the one with the imported ice bar and the DJ, the Puma party favors, and the near catastrophe with the collapsing tent.

But that was all before Moses's time, so Moses hadn't yet seen a full-throated Branson production with his own eyes, what Moses called "the Hollywood of it all"—the splendor and speaker stacks and lighting rigs suspended from the rafters. Moses felt that the party they were about to host was premature and would probably set them back a couple of weeks: they had rushed to button up the ship for the event. After everybody went home they would have to open it back up and finish putting in the wires and cables. But it was Branson's money, and he was the marketing maestro. If he thought a splashy event with celebrities and customers and lots of press was good for the business, then who was Moses to argue otherwise?

A fleet of black Mercedes vans pulled up bearing customers. Some had traveled thousands of miles. An engineer from Japan flew across the Pacific and said he had been dreaming of going to Saturn since he was thirteen and wasn't remotely deterred by the crash. "Until reaching my goal, I think many accidents may happen," said Taichi Yamazaki. That was Branson's magic, convincing six hundred people who had given him roughly a quarter of a million dollars ten years ago not to ask for their money back, that they were in for more than a suborbital space trip—to think of their investment like dues to an exclusive club that got them invited to Necker Island and fancy dinners in New York and a luxury escape in rural Idaho to watch the solar eclipse. They all got metal-plated business cards embossed with their names and their title: FUTURE ASTRONAUT. "I've already got what I paid for, so I'm just in for a bonus," said an Icelandic businessman.

It felt bizarre, almost like an invasion, for many of the techs and engi-neers. They were blue-collar guys whose colleague had just died at the bottom of a ravine. They weren't in the mood to host a rave for people they didn't know, wearing shirts made by a German designer that didn't quite fit because the Germans wore their clothes tight.

"I just tore my shirt," said one engineer.

Another was fussing with the hem, pulling it to try to stretch it out, and he wasn't even a big guy. He knew people who had asked for a double-extra-large or triple-extra-large and been told by the Germans, "We don't make them that big."

THE REPORTERS ARRIVED later in rental sedans and news trucks, the kind with a bedhead of antennas on the roof. They checked in at the reception desk with the candy bowl, waiting to be escorted upstairs, where, in the mission control room, Branson was sipping tea on a stool next to White-sides and Moses.

Many of these reporters had been on the space beat for years so they knew Branson and Whitesides—but from a distance, because Branson was a busy billionaire and Whitesides usually seemed stressed. It was different with Moses, who took the time to build rapport with people.

Moses told the reporters to feel free to record. Depending on what they wanted to know, he might have to get into details and, "If you know me from my NASA press conferences, I talk very, very fast."

He treated the press with respect. He supplied details if reporters asked, because if they asked, it showed they cared, whereas some of his peers, par-ticularly some who'd spent years working for Branson, patronized the press by trying to protect even minor details like they were sacred relics. I sat in one meeting where a Virgin employee boasted, "I know how to manhandle media." Later, when I went to the spaceport in New Mexico with Moses and the operations team, I listened to Jonathan Firth, a longtime loyalist who'd worked for Branson for twenty years, give his take of some recent media coverage—a pat to the Associated Press for their "mature view of things," a dart to the local paper for being "less positive" and always bringing up the crash when Firth kept telling them, "Don't want to talk about that any longer. That's in the past." He added, "We have to manage people's expectations." It was awkward hearing them talk about "narrative" in cynical terms; I couldn't help wondering whether I was the subject of similar meetings.

I took comfort knowing that all of the razzmatazz made Moses equally unsettled. He tolerated the fog machines but refused to put up with smoke and mirrors. Like when the PR team in London kept comparing SpaceShipTwo with the Space Shuttle. Moses didn't doubt their good intentions, likening Virgin's new spaceship to one people already knew and admired. But, no, Moses told them, comparing SpaceShipTwo and the Shuttle was disingenuous; they were vastly different machines with vastly different missions: the Shuttle went into orbit, while Virgin was just trying to get out of the atmosphere. He eventually had to pick up the phone and call London to tell them to stop. He made his point by giving them some hard numbers they could get their heads around. "Shuttle was going *Mach twenty-five*," he said. "We're going *Mach three*."

Branson took a question. The reporter asked about Virgin's longer-term ambitions. Branson said that flying people to space was "pretty cool," but, "Once you've got people into space, why shouldn't we have point-to-point travel at tremendous speeds? And why shouldn't we go on creating an orbital vehicle? We will start to do that. I just had a meeting with a senator, talking about asteroids. And they asked, 'Can Virgin Galactic come up with ideas to try to remove giant asteroids coming toward the Earth?' We'll have a look at that. And, 'Could Virgin Galactic help sort out the debris in space?' We'll have a look at that, too. And once all that's sorted we'd like to join the race for deep-space exploration."

You had to admire Branson's pluck and enthusiasm, thought Moses. But they were light-years away from achieving any of those things. When it was Moses's turn to speak he opted for more measured language. He talked about "evolutionary steps," which, in the parlance of flight test, he said, "we call envelope expansion."

He focused on the rigor of the process. "You've got a model on the ground that tells you what to expect. If the flight proves that your model's good, then you can trust its predictions for the next step. If it tells you it's not, you got to go back and do your homework before you can fly again."

Take the feather, he said. "The air tanks that hold the gas are in. They've been checked. We know they work. The plumbing that takes that gas out to the feather actuator is in. We've checked. We know that works. The actuator"—a pneumatic piston—"has been tested on the ground. We know that works. But we haven't sat in the cockpit and pulled the feather handle and watched it go up, right? We haven't done the integrated, end-to-end-type things. That's the next step."

In other words, they were still testing pneumatic pumps, he was saying: don't hold your breath for them to start zapping asteroids.

A wire service reporter asked the question everyone wanted to know: When were they going to fly?

Moses took this one. He hemmed and hawed and promised he wasn't "dancing," then said, "Schedule's one of those things. As soon as we tell you guys a date, we start being held to one and we don't want the team to feel that pressure." It sounded like a dig at Branson, constantly spouting dates.

Reading the room, Moses saw dissatisfaction. I know, he said, "I'm answering your question without answering your question."

Maybe Branson was right, that you could level with people only so much, that no matter how much people insisted on the truth, what they really wanted was entertainment and spectacle. And if that's what the people wanted, then that's what Richard Branson would give them.

WE FOUND OUR SEATS. Then, cued by tribal drumbeats and symphonic melodies, Stephen Attenborough took the stage. He gripped the podium so tightly that it rocked on its base. "Let me hear you make some noise!" he said, reading off a teleprompter.

Moses would eventually get a turn at the mic, but this was Branson and Attenborough's show, not his. Attenborough offered vapid words about the company's "higher purpose," which he called their "cultural due north . . .

uniting us in service of something bigger than ourselves." He thanked Land Rover for giving them cars and providing a "textbook example" of "the difference between sponsorship and partnership." The carmaker shared their passion for "the journey at least as much as the destination."

Then Attenborough played a pre-recorded message from the Nobel laureate Malala Yousafzai who said, "My superpower is to speak for girls in a voice so loud that the world will listen," and with whom Attenborough pledged to "definitely change the world for good."

Suddenly the curtain behind which Moses had slipped earlier parted to reveal SpaceShipTwo, towed by a white Land Rover. Branson, whom the *New York Times* once called a "one-man publicity circus," stood through the open sunroof, blowing kisses like an incumbent on the stump. People gasped amid dizzying strobes and pumping music. When the SUV stopped, Branson got out and marveled at the ship.

"Isn't she beautiful?" He recommended we all pinch ourselves.

The lights dimmed, and on the screens that had projected Yousafzai's image a single eyeball now appeared. The eyeball belonged to Stephen Hawking, the cosmologist and quantum physicist, whose work on the mysteries of time and the universe had done as much as anyone's to deepen our fascination with all that lies beyond. Branson recalled once hearing Hawking on the radio say that his dream was to fly to space, prompting Branson to call Hawking and offer him a seat on SpaceShipTwo. Hawking didn't know if his weak, ALS-addled body would survive the trip, but he once told Branson he was perfectly content dying in space, "surrounded by the stars."

The blue orb of Hawking's eyeball glowed above our heads in the otherwise black hangar. Then, hauntingly, his robotic voice came over the PA system.

"I have always dreamed of space flight," said Hawking. For most of his life he thought it was "just that . . . a dream." He had been confined to a wheelchair on Earth, left to experience "the majesty of space" solely through his imagination because space was "the domain of astronauts, the lucky few who get to experience the wonder and thrill." Now Branson wanted to make that wonder and thrill available to everyone.

Richard Branson at the SpaceShipTwo rollout, in 2016.

When the lights came back on, Branson brought his infant grand-daughter up. She had just turned one, and together he and she christened the craft by smashing a milk bottle on the ship's nose. Another guest, the soprano Sarah Brightman, stood and sang "Happy Birthday," and when it was all over waiters passed out flutes filled with champagne and blue curaçao while David Bowie's "Starman" played. Branson's PR team carouseled him in front of TV cameras for a shotgun series of interviews. He talked about building hotels in space.

When Moses did get his turn, he tried to shed some realism on the day. Yes, he said, today marked a "technical milestone," but he emphasized that there was "a whole lot of work" left to do. But even he knew that this audience didn't want to hear about that. They wanted razzmatazz and to hear about how they were going to change the world.

Or at least most of them did.

MARK PATTERSON, THE race car driver, skipped the rollout. His gut was bothering him. He had earned a lot of money in private equity over the years by trusting that gut, but now it was telling him something different: he was starting to doubt the wisdom of his investment. He'd always

thought of his ticket that way, not as a reservation but rather as an investment in an experience that only money could buy.

He used to attend the events with the other bright-eyed zealots who
cherished their membership cards and their SpaceShipTwo cuff links, and
their invites to Necker Island. It felt exclusive, and thus appealing. He saw
the logic and felt the momentum. He passed a physical exam and went to
train in the centrifuge, where he met another prospective astronaut, Stan-
ley Minkinow, a jeweler who had survived the Holocaust, but whom Virgin would disqualify because of his age. Minkinow objected, saying he
could handle it: he had been a former paratrooper in the US Army. At least
Minkinow got a chance to sit next to Buzz Aldrin one night at a Virgin
function. "That alone made the two hundred thousand dollars worth it,"
he said.

Patterson saw things differently. Now, instead of being in Mojave, he
was looking out at the incoming swells from the porch of his winter home
in Cabo. He was tired of Branson and Attenborough failing to deliver and
expecting Patterson to be satisfied. He was amazed by some of his fellow
customers. According to Branson, one customer told him not to worry
about space because "We're thoroughly enjoying spending all this time
going to the game reserve in Africa, or Necker Island."

Patterson hadn't paid for a safari. He was in it for a rocket ride, which
now seemed less and less likely. Being a Virgin Galactic "founder," Patterson would supposedly be on one of the first flights, but there were nearly
a hundred founders. Virgin could make ten or fifteen flights and he still
might not get his turn.

He had been thinking about that money and what else he could do
with it. Maybe he was wrong and would come to regret it. But this was
how he made his fortune, sizing up companies—betting long on the ones
he liked and short on the ones he thought would fail.

And as much as he wanted to climb into that ship and cinch the harness buckles and feel the weight of all that gravitational force pressing on
his chest as the spaceship broke through the atmosphere at three times the
speed of sound, he just couldn't see a path.

The company was bleeding cash with unrealistic revenue projections, based on some assumption that they would soon be airborne. But there was nothing certain about that assumption. And however much Patterson admired Branson's audacity, he couldn't see Virgin becoming a viable business. Better to get out now before it was too late.

A few months after the rollout, Patterson called Attenborough and asked for a refund. He got the money back right away and might have put it into some known, terrestrial venture. But he liked what he saw in the space industry. It was full of dreamers with big ideas. Some of those ideas were just better than others. And in the end? He invested in another rocket company, Radian Aerospace.

20.

JITTERS

STUCKY WAS WORKING LATE one night when he heard a *bang* on the other side of the hangar. He hustled over to the spaceship.

It was nearly midnight; Moses had teams working around the clock, stressing and testing the ship on the ground to ensure that it was ready to fly. Techs bent parts until they snapped, which felt sadistic and counter-productive to Stucky, but he didn't say anything until now, when he found a tech under the ship torquing his wrench on an actuator cable that had just snapped and made the *bang*.

Why were they were trying to break their own ship? Of course the cable would snap, Stucky said: "It's not *designed* for that!"

The engineers conceded that a wrench was unlikely to get lodged on an actuator cable mid-flight. But the exercise proved that the cable was too weak. Now they had to build a bigger and stronger one. It would inevitably be heavier than the old one: every pound would cost them in performance.

These were more than engineering decisions. They were philosophical ones, statements of their tolerance for risk. Stucky didn't like what they said. He feared that some engineers were drawing the wrong lessons from the crash, letting fear dominate their calculations. Everybody wanted to build a safer ship, but Stucky worried that risk aversion would compel them into building a tank with wings that fell from the sky like an anvil. They could never fully eliminate risk; it was marbled into life, like sirloin fat. We each made our own calculations. Sometimes smokers wore bicycle helmets.

Increasingly, Stucky reminded colleagues that they were doing something incredibly risky. If they couldn't live with some risk, then they would never roll the ship out of the hangar, never get to do the thing they came to do. He had seen it happen to otherwise excellent engineers and saw it happening again now.

A few honest ones admitted that they were fighting doubt like an undertow dragging them out to sea. "I was becoming more risk averse," said one engineer, Matt Kampner. "We as engineers often don't have the ability to assess the big picture. As humans we are notoriously bad at risk assessment and probability. You have to consciously make an effort to do this kind of analysis."

Kampner had spent eight years working on SpaceShipTwo and WhiteKnightTwo and had learned many things, but perhaps above all it was that test pilots were just wired differently from the rest of us. It was as if they were programmed to constantly be assessing and reassessing risk. "Normal people don't do that. Convincing a fresh-out-of-school engineer responsible for the wing to figure out how many plies to use? The weight on their shoulders is just unimaginable. They clam up. They get really conservative. They can't accept *any* ambiguity," said Kampner.

The crash exacerbated those tendencies. Kampner had colleagues coming straight out of school or from other industries, like vacuum cleaners, where nobody's life was on the line. "They were scared. I felt that," he said. He was, too.

At one point, he and Stucky clashed over the landing gear. Kampner wanted to redesign the gear, which had been made for a lighter spaceship. He, and others, said the margins were thin, by FAA standards, and that they were lucky the gear on the other spaceship hadn't buckled on them as they were landing, sending the craft skidding off the runway like a greyhound with a broken leg.

Stucky countered and questioned the basis of their margin analysis. He pointed out that holding SpaceShipTwo to FAA standards was foolish because those standards covered every condition imaginable and the spaceship wasn't even going to fly in the rain; asked Kampner whether he had thought through all of the downstream modifications the gear redesign might require, like whether it shifted the center of gravity or put undue stress on the wing spars; and, finally, contended that the design they used on the old ship was plenty rugged for their needs. He would show Kampner and he could see for himself.

The landing gear was boxed up in an unmarked hangar with the rest of the wreckage. Nobody was supposed to go in there, for liability reasons. But Stucky had a key, and a moral compass, and he dug through crates full of debris until he found the gear. He took an engineer who used a coordinate-measuring machine to scan the part, and at a meeting the next day he presented the data that proved his point—that the gear was as good as new after more than thirty landings and one catastrophic crash.

He eventually pulled Kampner aside, not to bully him or gloat, but to remind Kampner that all he could do was analyze, assess, and remember that they could never eliminate every risk. "It wasn't him telling me to stop being such a pussy or anything," said Kampner. "More him saying, 'Be reasonable.' And he was right."

But it was too late. Kampner had decided on a lower-stakes industry and soon moved back home to Pittsburgh, to pursue a career building robots.

TWO MONTHS LATER, the sun was shining and Stucky walked out of the hangar onto the flight deck. SpaceShipTwo was mated to WhiteKnightTwo, ready to go. It was the morning of September 7, 2016, twenty-two months after the crash. Finally, they were about to fly.

Stucky circumambulated the ship, homage masquerading as inspection. The Mylar on the tail booms caught the sun, and Stucky joked with the maintenance crew about them heating their lunches in the reflection. Stucky was palming the side of the ship, like a jockey patting his horse, when another pilot came over in a huff.

"We don't have any paperwork to fly!" said Todd Ericson.

Ericson had blue eyes and rosy cheeks that flushed and flared when he

Todd Ericson (left) at the crash site.

got angry. He'd always been the boy wonder, with his neat hair begging to be mussed and the roomy leather jacket he wore that could have been his father's or his brother's or one of those items bariatric surgery patients hold up after the procedure to show how much weight they'd lost. Ericson had been a prodigy at the Air Force Academy, the star of the flying team who aced a national flying contest with a perfect score. He got commissioned, flew F-16s, and rose through the ranks.

Ericson got to know Stucky at the end of his career, when Ericson took command of a top secret squadron in Nevada. Stucky was out on the range doing classified work for Scaled. Stucky liked him; Ericson had the same kind of knowing, aggressive smirk that he did. Let him into the cockpit and he'd figure it out. This was what drove Ericson crazy about Virgin: every test program had phases and they had spent long enough in the design phase and now they were in the test phase and that meant it was time to go, that they needed to get into the sky and figure out what was working and what wasn't and make the best of what they had.

"Better is the enemy of good and right now is she good enough?" Ericson asked. He wasn't worried about all the things the ship couldn't do. "Tell me what it *can* do and then we can assess whether we can live with that or not," he told the engineers. Stucky trusted competent people with confidence. "Giddyup!" Ericson liked to say.

Now Ericson was fuming.

They were supposed to fly in fewer than twenty-four hours. It was a modest mission, what they called a "captive carry," whereby White-KnightTwo would take off with SpaceShipTwo latched to its belly, and the pair would fly around for a couple of hours so the engineers could gather data on how the air passed over and under and around the spaceship. A live wind tunnel. It wasn't particularly risky, but you could never know. What if the latch failed and the spaceship suddenly dropped from the belly of the mothership like a bomb? Things happened. In 1953, when the Air Force did a captive carry test over the Great Lakes with a rocket ship hanging from the bomb bay of a strategic bomber, the rocket ship suddenly exploded and dropped, leaving a fiery hole in the belly of the bomber that instantly killed the test pilot and a crewman who was observing from the mothership. They had to be ready in case something like that happened. But they weren't ready; they didn't even have the right paperwork.

Ericson and Stucky marched into the hangar, ready to grab someone by the lapels. That was a problem around Virgin. Somebody should get an earful for late paperwork but nobody ever did. What happened to the "fine art of chewing ass"? Ericson wondered. He and Stucky charged into a small conference room, its window hung with tattered, eggshell mini-blinds.

"What the hell?" said Ericson. He directed his criticism at Doug Shane, who had hired Stucky at Scaled and was now the president of Virgin's manufacturing subsidiary, The Spaceship Company. "To be here, five hours before the first brief, and to be having this discussion? We have failed. This is not the way to run a flight test organization," said Ericson.

Shane leaned back in his chair with his hands clasped on his head. Was there a difference between nonchalance and flippancy? "We can all be frustrated," said Shane. "But we are where we are. And we're moving forward, not backward."

I was sitting behind Shane, against the wall. He looked over his shoulder. No one liked being criticized in public, even if it was in private. "I hope you realize that you're sitting in the middle of a very important discussion, and we're being very open and honest and I don't want anyone in this room to feel like this is going to come back to bite us," Shane said to me. "Are we clear on that?"

I asked what he meant. Clear on what?

"I feel uncomfortable with you sitting in here and I'm trusting that this won't turn into something that we'll all regret later," he said. A pause that contained an invitation to leave.

I thanked Shane for his trust and kept taking notes and hoped that when I looked up he wouldn't still be looking at me.

Stucky cleared his throat. "Doug made a point," he said. "'Let's not talk about this. We're *here*.' And we are here." By *here* he meant now and in this room but he meant more than that, he meant behind schedule, disorganized, and waiting on things to be done that should have already been done. "But maybe we don't always need to be here. It seems like we are always here."

He suggested a poll, a yea or nay, about who felt ready to fly tomorrow if they could get the paperwork finished in the next few minutes, as they were hoping to do. Nods all around, but Stucky wanted to double-check with Mark Bassett, the project engineer.

"McLovin, what do you think?"

"There are some 'unknown unknowns,'" said Bassett. "But those are relatively small risks."

"Are you concerned?" asked Stucky.

Bassett shook his head.

"Then neither am I."

AT THE PREFLIGHT brief the next morning at five thirty a.m., Stucky was jotting notes in his logbook when a maintainer broke the bad news. He was just out inspecting the ship with a flashlight when he noticed a one-inch hairline crack along the spine where SpaceShipTwo attached to White-KnightTwo. Stucky looked up from his logbook.

"No pressure," said Moses. "If it's not good, we don't go."

To go or not to go, a decision so plain and simple, but one that kept launch managers reliably awake at night because so much depended on getting it right. In the days of Mercury and Apollo, NASA's mission control room was full of men with clammy hands who knew that "a stricken instant, a cauldron of adrenaline, a failure of nerve" could "lay a shadow upon all the hours of his life" if they unnecessarily aborted a mission or proceeded with one that "resulted in death."

Moses told Bassett to analyze the crack, to see whether it was prone to splinter and spread or whether it seemed contained.

Bassett left. Stucky stayed in his seat for a moment, confused. How many people had inspected that ship during the last few days and not seen a thing? Was it an old crack? Or a new one? It was weird, whatever it was. He took comfort knowing that Bassett was on the case. Stucky trusted that Bassett wouldn't let emotions color his analysis, that he understood the difference between a real risk and an imagined one.

They broke for breakfast.

Bassett returned with his verdict ten minutes after everyone was back in their seats.

"Let's start with the good news," said Bassett. Because of the crack's location "its ability to propagate is very small." But he didn't know what caused it or how long it had been there. And though engineers tended to despise mysteries, Bassett had ruled out the worst possible mysteries and therefore didn't think this one justified scrubbing the flight.

Stucky felt relieved.

But then Ericson polled the room. "Any dissenting opinions?" he asked.

A hand went up. It belonged to Wes Persall, a flight-test engineer with spiked hair. "I can't lie and say I'm not a little hesitant," said Persall. He had been with the program for years and been friends with Mike Alsbury and happened to be sitting in the back of the spaceship when Stucky got into the inverted spin on glide flight sixteen—the one and only time they put an engineer in the back. "We've come a long way, and it would be a shame to do something now that would delay us even further," said Persall.

WhiteKnightTwo and SpaceShipTwo flying a "captive carry" mission.

Stucky saw heads nodding in agreement, and he wanted Moses to do something soon before everyone raised their hand and said they were nervous.

But Moses let it play out. Besides, he was going through his own "jitters." He had recently been in the office alone one weekend afternoon when he found an old photograph of Stucky and Alsbury, taken after the first powered flight. They were smiling and gripping hands. Moses lost it and broke down at his desk. He knew why, but then again he didn't. "A man died, and his family was torn apart," he said. But at the same time that man's mistake had cost Virgin dearly; if that hadn't happened, Moses thought Virgin might already be flying tourists into space. "Then you feel guilty that you feel just as bad about the program as you do about the people most affected. So you bury all this, because it doesn't seem right to feel more, or as, sad about a thing as about a person." He had wrestled with similar feelings after the *Columbia* crash: "When I see a picture of the *Columbia* crew, I get a sadness, a feeling of great personal loss. But I also can't not remember the six months we spent cleaning up debris in the forest. And that's a whole different kind of hurt—a realization that the ship was gone and that the program was over."

Persall said he had watched enough training videos about risk and accident avoidance to know that he should listen to his gut.

Finally, Stucky spoke up. He had seen all those same videos, he said. But he drew a different lesson from them. "What you don't see in those videos is when they have these kinds of meetings and then they go up and nothing happens," he said, and with experimental vehicles this kind of thing happened on "practically every flight."

Persall gave in and they flew. It all went fine.

Still, something wasn't sitting right with Stucky; something about Ericson opening the floor like that to questions made Stucky uneasy. It shook his trust. What was Ericson thinking? That wasn't the "Giddyup!" side of Ericson that Stucky knew. Ericson seemed less like a man on a mission and more like one making choices for posterity—as if Ericson was intent on casting himself as the scrupulous one, in case of another postaccident report yet to be written.

21.

THE WALNUT

STUCKY DROVE INTO THE desert on Halloween. It was dark when he left the house. It was always dark when you thought about Halloween, and the spooky stuff usually happened at night. The crash proved that awful things could happen during the day, too.

Stucky reached the crash site as the sun was coming up. A handful of former Scaled colleagues met him there, friends of Alsbury's convening at dawn to commemorate the second anniversary of his death. They brought a bouquet of white roses and passed them out. It was cold: they wore their hoods up and their sleeves cuffed over hands gripping flower stems.

"Is this appropriate?" an engineer named Matt Stinemetze asked Stucky. He was waving a copy of John Gillespie Magee Jr.'s poem "High Flight."

Of course, said Stucky.

Stinemetze read the poem aloud—

> Oh! I have slipped the surly bonds of Earth
> And danced the skies on laughter-silvered wings;
> Sunward I've climbed, and joined the tumbling mirth
> Of sun-split clouds,—and done a hundred things
> You have not dreamed of—wheeled and soared and swung
> High in the sunlit silence. Hov'ring there,

I've chased the shouting wind along, and flung
My eager craft through footless halls of air . . .
Up, up the long delirious burning blue
I've topped the wind-swept heights with easy grace
Where never lark, or eagle flew—
And while with silent lifting mind I've trod
The high untrespassed sanctity of space,
Put out my hand, and touched the face of God.

—and when he finished, he and Stucky and the others laid their roses in the dirt, returned to their cars, and drove back to Mojave, like it was any other day.

But it wasn't any other day. Stucky was scheduled to fly a glide test the next day. Some might have said: No, it's too close, it doesn't feel right. But none of that bothered Stucky. He didn't believe in omens or auspicious days. Maybe that's what irked him about that poem, with its line about touching the face of God. Because when you started conjuring and relying on God, you stopped relying on science and that's when things got dangerous.

He was religious about the weather. Pilots sort of had to be. "The love of flying demands the attention of a lover to the moods of weather," said Norman Mailer. Stucky's attention bordered on fetish. He curated forecasts, scoffing at simpletons who relied on any one single model. He preferred to consult them all—the ICON, the NAM, the GEM, and so on. He got excited about solar density data, lapse rates, and discussions about "arcane micrometeorological matters."

His concern about tomorrow was about the weather. None of the models were looking good. All showed clouds and bluster.

Already he felt the wind picking up: it was pushing his car all over the road on the drive back to Mojave. He might have spent the rest of the day worrying about it, and then lain in bed all night, hoping a gust wouldn't catch the broadside of the spaceship on his landing approach and knock him off course. He might have, but why? It was no use worrying about things beyond your control.

Focus. Attend to the job at hand.

That's what they taught him and that's what he trained himself to do. All the elite performers did it—the top pilots and quarterbacks and three-point shooters, the ones who always pulled through in the clutch. They tuned out the noise. They compartmentalized distractions. They found "silent sanctuary" in the cockpit, on the court, in the pool, on the field. They suspended their emotions, even illnesses, to perform. It was uncanny. Michael Jordan played through the flu, scoring thirty-eight points and leading his team to an NBA Finals victory. Michael Phelps swam and won "independently of his feelings," said his long-time coach. NASA astronaut Mark Kelly led a mission into space four months after his wife narrowly survived an assassination attempt by a deranged gunman who killed six others, including a nine-year-old child.

But then again, we were all flawed in our own ways, and something *had* been bothering Stucky: his pet Chihuahuas. He and Agin had two of them, Igor and Spike. They had recently gone on vacation and put the dogs in a kennel—a fancy operation with pickup and drop service, and a menu of add-ons. Stucky ticked the box for behavior training.

They went away and came back, but Carlos the kennel owner said he needed more time.

Enough is enough, said Stucky. Give me my dogs back.

Sorry, said Carlos; Igor was at the vet: failing liver. And now he needed a few more days to finish Spike's behavioral regimen. Then Carlos stopped responding to their messages. He cited a car accident. Agin rushed to the vet, but Igor soon died. Agin grew terribly upset.

Stucky demanded that Carlos give him an address. When Stucky got there he found an empty lot. As far as Stucky was concerned, Carlos had kidnapped his dog.

Turning sleuth, Stucky clicked on a photo Carlos had sent of Spike posing before a hay bundle. Stucky retrieved the metadata and plugged the coordinates into his phone. He arrived at a rural address and, ignoring the posted guard dog warnings, rumbled up the gravel driveway to a big house and an adjoining fence-work kennel with a yelping scrum of dogs.

Stucky stepped out of the car. "Some meth head is going to shoot me," he texted Agin. A big dog with prominent incisors charged him but was jerked back by its leash.

Stucky saw Spike among the scrum. He approached the fence, pulled it back, tucked the Chihuahua under his arm, and ran. He jumped in his car and sped away.

"I just broke into private property and stole the dog," he told Agin.

"What the hell?"

He thought she would be happy but he got an earful instead. Now he was trying to put all of that into a box, a compartment in his mind, so the next morning when he climbed into the cockpit, his silent sanctuary, he wouldn't have to think about it and could focus on flying.

AN HOUR BEFORE takeoff, Mike Moses was sweating the wind. He sat in the overflow room next to mission control popping his phone from the holster on his hip, checking the latest wind readings, seeing that they didn't look good, holstering his phone, and then doing it again.

He was accustomed to fickle weather. Summer thunderstorms used to make NASA launches terribly unpredictable. "People make a fair bit of fun about my last name being Moses," he would say. "Like I ought to be able to control the weather, but I just can't, so we'll deal with what we get."

He hoped for good weather but you often learned more about people in bad weather; adversity revealed character, testing nerves and mettle, telling you with whom you'd want to share a foxhole. Moses was always looking for the next test, ways to stress the ship and the pilots and the engineers to the verge of breaking so he could identify the edges of the envelope.

Did anyone know SpaceShipTwo's headwind limits off the top of their head, he asked?

Twenty knots, came a reply.

He knew that. They all knew that. He was just testing them.

They weren't supposed to fly in winds greater than twenty knots. It had been gusting well above that all night, but now it was blowing eighteen, twenty, seventeen, twenty-two, nineteen knots—flirting with the limit. One of the forecast models predicted a lull in about an hour. Maybe they could take off, time the lull, release the ship, glide down, let Stucky hit all the test points, and then land before the gusts picked up again.

"We better get going," said Moses, grabbing a walkie-talkie on his way out. I followed him and three engineers into the parking lot.

We squeezed into Moses's SUV. It was cold and smelled like auto shop coffee. Moses offered his daughters' blankets in the far back if anyone wished to bundle up. "You want the Winnie-the-Pooh blanket?" he asked. "Or the heart blanket?"

Moses drove to the other side of the airport where he showed his badge to a security guard who promptly lifted the boom and waved us through. Cloud globs hung low in a Picassoesque blue sky.

Moses parked and zipped on his jacket. People had braved the cold to come out to watch. You could see their breath, and their infants tucked in down cocoons. Beth Moses had UNITY, the spaceship's official name, stenciled on one cheek.

The airplanes took off but had no sooner begun to climb when the control tower reported a twenty-three-knot gust. Had Moses pushed too hard? Was he going to regret this?

"This is insane," Doug Shane, the easygoing president of The Spaceship Company, said to Moses. "Look at that wind sock." It was bright orange and straight as a scolding finger.

Another gust came in at twenty-one. Moses winced, but he knew what had to happen: the flight was aborted.

"Knock it off," he said, over the radio.

Later, when he gathered the team back in the briefing room, he tried to put on a positive face to address their frustration. He obviously wished they could have flown and gotten that one out of the way and been one step closer to resuming their rocket-powered program, which was what they all came to do. But flight test was about facing the unknowns, he said, and, "sometimes the only way to see those unknowns is to go out and fly."

THEY COMPLETED THE glide flight a month later and squeezed in another just before Christmas 2016. They carried that momentum into the new year, marching through a procession of glide tests. An earlier plan called for twelve glides, but Stucky scrutinized the plan and thought if they could combine the test points they could do what they needed to do in seven flights.

Stucky got into fighting shape. He went on a diet and attended some yoga and kickboxing classes. He sharpened his stick-and-rudder skills by flying sailplanes, his paraglider, the simulator, and WhiteKnightTwo. Virgin also owned an Extra 300L, a nifty twin-seat aerobatic prop plane that Stucky could take out whenever he wanted to roll and flip and bank to the verge of passing out in order to build up his g-tolerance.

G's, or gravitation forces, came in three categories, based on how they hit you. You felt g-x sagittally, like a bullet in the heart; g-y laterally; and g-z vertically. The g-z were arguably the worst because of the blood rushing to and from the brain. Based on the engineering models and medical experts, SpaceShipTwo pilots would feel about four g-x on the way up and five g-z on the way down. Five g-z could be a bit unpleasant so the engineers were designing a special seat for the passengers that would recline upon reentry in order to shift some of the g-z to g-x and limit the g-z to four, a manageable amount, the experts believed.

Stucky loved pulling g's, feeling the pressure build in his head and on his chest, flooding his brain with dopamine. It felt like almost a crime, getting to fly acrobatics on the clock, but it made sense—it was part of the regimen, conditioning his mind and body to tolerate levels of stress that would freak out ordinary people. That's who you wanted flying your spaceships, pilots with compartmented brains and calloused souls who bent without breaking.

One day Stucky offered to take me up. I agreed. He led me to his desk and tore off a yellow sticky note. "They want you to write down next of kin," he said. I scribbled my wife's phone number.

We changed in the locker room, emptying our pockets and putting on harnesses. A doctor offered some suggestions to avoid getting sick. Everybody thought they knew a trick. Before Branson went up he drank a cup of prune juice because he remembered his mother saying prunes were good for upset stomachs, when in fact they were good for constipation; "Mum was dyslexic," he said.

It's all about core and the anus, said the doctor, Tarah Castleberry. When the g's start coming on, she said, "pretend like you're squeezing a walnut down there."

We climbed into our seats and sealed the canopy. I tucked a sick bag under my leg. Stucky kicked over the propeller and it whirred to life, sounding like an amplified wasp.

We took off and made a steep, corkscrewing ascent. At about 8,000 feet, Stucky tilted one wing to look down and confirm there weren't other aircraft below, then jerked the stick and banked to one side, pulling five g's. He held this for fifteen seconds. Then he pulled back like he was doing a loop but he stopped halfway and kept the airplane upside down so we were out of our seats, feeling negative g's. He held this for fifteen seconds. Then he flipped the Extra right side up and banked into another five-g turn. I reached for the sick bag to ensure it was handy. He held this for fifteen seconds.

Everything served a purpose, an offering at the altar of envelope expansion. The experts had plugged in data and consulted computer models but couldn't say for sure how the human body would react on a SpaceShipTwo flight. "Usually when you go to space you see very high g"—during boost—"followed by long periods of zero g"—in space—"before you return to gravity. You're up in space for at least a day or two, if not weeks or months," said Moses. Scientists knew what happened to humans during space flight—elevated intracranial pressure, swollen optic nerve sheaths, flattened eyeballs, chromosomal changes, blood clots in the jugular vein. But SpaceShipTwo was unique because of its initial high g, followed by four minutes of zero g, followed by more high g. It condensed the timeline. "There's no scientific data to show what will happen," said Moses. Stucky tried to mimic the singular elements of this profile in the Extra—the high g's, the negative g's, the high g's.

He asked how I was feeling.

I said that during that initial five-g bank my vision had temporarily grayed out and my helmet felt pinned against the headrest, like it was being swallowed by quicksand, but when Stucky pulled out of the turn the sensations passed.

He laughed. The quicksand sensation was common, he said—that feeling of some invisible, superhuman force holding you down. He told me a story that his RIO used to tell about being in the back seat of an F-4 during air-to-air combat training, and bending over to look through the radar scope between his knees just as the pilot yanked on the stick and got into firing position.

"Lock him up!" said the pilot, and the RIO was trying to do just that, to hit the radar lock switch, but they were still pulling g's and the RIO couldn't lift his head up from between his legs. "I can't!" the RIO shouted. "I'm giving myself a blow job." It still made Stucky laugh, all these years later.

Honestly, I was feeling okay until Stucky said, "This is something I saw on YouTube."

He rolled us upside down, pulled the stick in his lap, did three-quarters of a loop, then pushed the stick forward. We began tumbling toward the Earth, nose down. He held the dizzying maneuver for several seconds.

I conjured the walnut.

WHEN I WAS young I dreamed of becoming a fighter pilot. I had a flight suit that a tailor in Seoul had made specially for my dad. It was green, just like his, but with my name embroidered in yellow across the breast, and I would step into the legs and drag that long zipper up the front and feel like I was putting on another skin.

And, always, airplanes everywhere—jets roaring overhead, and balsa wood models in the garage, and remote-control planes that we would open on Christmas morning and inevitably crash into a tree before lunch. My brother and my dad and I took hang-gliding lessons on the sand dunes of Kitty Hawk. You can get your pilot's license when you turn sixteen, my dad said. He knew people with antique biplanes and gliders and a single-prop Piper Cub who took me flying and offered to get me qualified.

My dad had views on adolescence, on fear, on character, on how boys became men, on rites of passage. Years later he told me, "I was trying to create some kind of climate in which you could grow." We hunted wild boars in the Lowcountry, our faces smeared with camo paint, paddling

From left: Pamela Schmidle, the author, Robert Schmidle.

through the creeks and then dropping anchor in the mud to get out and stalk boars with tusks curved like fishing hooks, our boots tied extra tight so when we sunk into the mud we didn't lose a shoe. When I was thirteen I shot and killed my first boar, boiling its head in a pasta pot so I could remove the teeth and make a necklace with the curvy tusks.

When I was sixteen we moved to Virginia. I was moody and into books about solitary adventurers. One day my parents drove me to the Blue Ridge Mountains and left me on the side of the road, near the Appalachian Trail. I went off and hiked alone, me and my moods and the imaginary sound of black bears invading my camp in the middle of the night. I met my parents the next day twenty miles up the trail; I knew my dad was proud.

More rites, more passages.

After I graduated from high school my parents sent me to Wyoming for a week. They hired a rock-climbing guide to lead me up crags and cracks on daunting routes in the Wind River Range. It had to be a private guide

The author.

because my dad believed that you diluted an experience when you did it with a group; my earlier bid to join the Boy Scouts never stood a chance.

I think my dad assumed that I would arrive where he did, that I would see flight as the ultimate transcendent experience, this opportunity to shed what Charles Lindbergh described as "all conscious connection with the past," to live "only in the moment . . . crowded with beauty, pierced with danger." But I lost interest as I grew up. I no longer dreamed of becoming a fighter pilot. I can't explain why. Maybe I knew how much my dad valued commitment, how he didn't believe in accidental excellence, and perhaps I was scared—not of crashing and dying but that I might not make it as far as he did, that I would wash out of boot camp or flight school, or end up flying cargo planes or helicopters or anything other than fighter jets, because I grew up hearing him talk about the difference between fighter pilots and everybody else, and I didn't want to be everybody else.

I went off to college and didn't turn back. Or at least that's what I told myself, that I would do things my way by studying philosophy, even though that way was already his way.

We live enveloped in whale lines.

It took more than a decade for me to realize that I never really wanted to become a pilot so much as I wanted to become like my dad, to achieve what others deemed impossible. Time alone did not prompt this revelation; having sons of my own did.

Gradually, I saw some of my dad's flaws recast as strengths. I realized what he was doing when he was not home, like when I read the letter he sent to his astronaut friend, Sonny Carter, describing those bombing raids into Iraq, and it dawned on me that he was my age when he wrote that letter, and that I had just spent my morning pruning hydrangeas in the backyard, which made me wistful for the awe I used to feel as a toddler when I would put on my father's flight helmet, and anxious over how I could possibly instill my boys with that same degree of awe.

He had performed feats that I could not imagine performing. I wanted to know how he felt and what he was thinking at the time. But my dad was protective of his inner life. He talked, but seldom shared; he set out to shape me and my brother, who went into the Marines and became a special operations officer, by modeling what he considered desirable behaviors rather than relating with undesirable ones. He never sat me down when I was a teenager to say that he knew what I was going through. "That's not a conversation anybody ever had with me," he explained.

Perhaps this was what attracted me to Stucky: his openness about his own flaws and fallibilities. Or maybe it was because Stucky was a stranger. I could ask him about his feelings in a way that I could never do with my dad. Through Stucky, it seemed, I could rummage vicariously into my father's inner life, to try to learn something about him, to try to figure out what it was that made him do what he did.

STUCKY PULLED OUT of the maneuver, applied power, eased back on the stick, and, just like that, we were flying straight and under control. I released the walnut.

He later described the tumbling sequence as being "sort of like a Lomcovák"—an acrobatic maneuver that means "profound shaking" in Czech. He did Lomcováks all the time but was always looking to mix it up, to see what other people were posting on YouTube. Doing the same routine at the gym every day gets you only so far.

And while the Extra was thrilling to fly, it had its limitations: it wasn't particularly fast and the cockpit didn't bear any resemblance to the one in SpaceShipTwo. This was part of what made flying the spaceship so hard: you couldn't mimic the experience in training. It wasn't like a race car you could drive around the track in second gear before you were ready to stomp on the gas and start the race. Combustion was all or nothing.

Stucky and the other pilots tried to work with what they had. They would rehearse the flight profile in the simulator, whose cockpit mirrored the one in SpaceShipTwo, but the simulator offered none of the physical sensations of flying under boost. They had to train in a centrifuge for that.

Virgin didn't have a centrifuge. These were hulking and expensive machines. The military had some, but the nearest one Stucky could use was outside of Philadelphia, three thousand miles away. So in March 2017, as the company advanced through the glide-flight portion of the program, and got ever closer to resuming powered flight, Stucky and two other pilots flew across the country to get spun.

22.

GRADATIM FEROCITER

STUCKY WAS EATING RUNNY eggs from the hotel buffet when his colleague Mike Masucci, the U-2 pilot, joined him at the table.

Masucci was a tall, loping man with a Roman nose and D-cell fingers.

Stucky asked him how he slept.

Both had been spun in the centrifuge the day before. Stucky was a bit sore. Over his career he had gotten spun more times than he cared to remember, but he was young and sprightly then—in his physical prime. Now, though his mind felt as sharp as ever, his body was slowly breaking down. He had achy knees and deteriorating eyesight. On a recent weekend afternoon he had put on surgical gloves and drilled pinholes into the visor of his flight helmet, then used a needle and thread to stitch a pair of reading lenses into the visor; the latex gloves reduced smudging. He felt his days were numbered. This might be his last hurrah.

Masucci said his legs had broken out in g-measles when he got back to his room, but the g-measles went away, and now he was ready to get back in the 'fuge, as he called it.

Others were supposed to join them, but Rick Sturckow, the former NASA astronaut, had gotten delayed leaving California, and Dave Mackay, the Scot, had fallen off his bike a few days earlier, breaking his collarbone. When Stucky found out about Mackay's injury he messaged Moses

and pledged to avoid any "potentially hazardous outside activities" until Mackay was back on his feet.

"I'm not sure your definition of 'potentially hazardous' matches normal definitions," said Moses.

Fair point, Stucky replied: "Somebody that eagerly looks forward to lighting an experimental rocket motor on an experimental plastic spaceship without an ejection seat and spacesuit probably doesn't fantasize about needlepoint on their off days."

Sturckow appeared across the lobby. He had on Levi's, work boots, a plaid shirt, and the red Marine Corps baseball cap that he was rarely seen without. He was wheeling a suitcase.

Stucky looked up. He deadpanned, "I don't believe I said, 'red hats.'"

Sturckow ignored him. He was famously curt; his call sign, "C.J.," was short for "Caustic Junior." But he and Stucky had history—in the Marines, at NASA, and now with Virgin—and although, or maybe because, so many people gave Sturckow such a wide berth, Stucky found a certain joy, almost a duty, in antagonizing him.

Despite the similarity of their names the two men were different as could be. Sturckow wasn't a dreamer. When he was a kid, he drove tractors and tended to turkeys on his parents' ranch and later worked as a lube boy for a big ag company. He grew up wanting to be a diesel truck mechanic. He studied mechanical engineering at Cal Poly and fell in with some gearheads who raced off-road; they would leave campus on Fridays after class and drive all night, sometimes as far as the Baja Peninsula, to make the starting gun on Saturday morning. Space flight never crossed his mind.

A professor urged Sturckow to consider the military, and he did. In 1984, he joined the Marines. He earned his wings and got orders to Beaufort, South Carolina—into my dad's squadron.

Sturckow was the odd man out, a straitlaced teetotaler in an otherwise rowdy and hard-drinking squadron. But he was incredibly smart and unmistakably talented. (He and his wife lived at the end of our street; she used to babysit for my brother and me.) My dad wrote Sturckow's recommendation for Top Gun.

In 1989, Sturckow led the recovery effort in Japan for my dad after the

C.J. Sturckow, in 2009.

flight control system failed on my dad's F-18; my dad tried to save the plane but it was spiraling into a mountainside when he finally pulled the ejection handle that blew off the canopy and ignited the rocket under his seat that, *thwap*, shot him into the air while his jet fireballed into the side of the mountain; he floated down under a parachute and landed in a tree.

Two years later, Sturckow was my dad's wingman on the first night of the Gulf War, jinking left and right to dodge incoming SAMs. Sturckow took a lot of flak for being all business, but when it came down to business, he was who you wanted on the job.

After Sturckow got back from the war he went to NASA, where he met Mike Moses and flew the Space Shuttle. Sturckow was an odd fit there, too: when a veteran NASA reporter made a list of superlatives to commemorate twenty-five years of the Space Shuttle program he named Sturckow the second-strangest astronaut to ever put on the puffy suit. But Moses had always liked him and trusted him; he thought of him as one of those perennially misunderstood types. Later, after Moses got to Virgin and saw Sturckow's name on a short list of prospective hires he hesitated for a moment, because he didn't know how Sturckow would fit with the culture, but then decided that Virgin probably needed someone with Sturckow's austere and single-minded focus. Sturckow got the job, though it wasn't clear whether he ever got the culture.

He left his roller board at the table and went off to peruse the buffet.

Bewildered, Masucci looked at Stucky, and then at Sturckow's suitcase, and then back at Stucky. He just got here, said Masucci. Was he checking out already?

"Who knows?" said Stucky.

"Bet he doesn't trust the hotel staff," said Masucci.

Sturckow drank a glass of orange juice. He asked whether Stucky and Masucci were planning to put on their flight suits.

Stucky shook his head. He explained how they had worn their suits the day before because someone said a photographer might come around and they wanted to look professional. But the photographer never showed up and they felt silly and overdressed wearing their suits with sneakers. "Something about wearing a flight suit with tennis shoes makes me feel like Joey Bag o' Donuts," said Stucky. Today he was wearing jeans.

Exactly, said Masucci. A flight suit was meant to be worn in an airplane with combat boots. Pairing one with sneakers reminded Masucci of the frail, tenderfooted airmen in boot camp whose feet would blister and who got medical waivers that allowed them to wear sneakers. He'd associated the tennis-shoes-and-flight-suit combo with weakness ever since.

Masucci checked his watch and saw that they were due at the centrifuge soon. "Let's get going," he said.

WE DROVE SEPARATE cars and Stucky rode with me, navigating us through an industrial park off the highway, north of Philadelphia, to a warehouse with a sign for the National Aerospace Training and Research Center, or NASTAR. We registered at the front desk and were escorted down a hallway decorated with portraits of Alan Shepard, John Glenn, and the Apollo 11 crew—PIONEERS OF SPACE, a placard read.

NASTAR was a vast facility sprawling over nearly half an acre. Richard Leland, the NASTAR president, gave a quick tour of the workshop where his engineers designed air-pressured seats, virtual-reality glasses, and decompression chambers. Leland had a cane and a dark sense of humor and told an awful story about once witnessing a test pilot suffer a stroke in a decompression chamber, and how the pilot emerged from the chamber looking "like he had a bad case of cerebral palsy."

Stucky grimaced.

"On that happy note," said Leland. He led the pilots up a flight of stairs, into a ready room with leather couches and flat-screens and a big window that overlooked the centrifuge.

Sturckow went to the window and stared at what looked like a gargantuan board-game spinner—this seventy-five-ton contraption, with its twenty-five-foot steel arm that could spin more than sixty miles an hour and create more than nine g's of gravitational loads on the one-man capsule at the end of the arm. It was a major advancement over the stone wheel that Charles Darwin's grandfather used to devise the first human centrifuge in the late eighteenth century: he'd envisioned tying subjects to a corn mill to test whether the steady rotations would trigger sleep.

A tech powered up the centrifuge, initiating its hypnotic routine. "It's kind of making me nauseous just looking at it," said Sturckow.

He was up first. Tarah Castleberry, the doctor, checked his vitals and strapped a neoprene band around his chest to record his biometrics. She followed Sturckow downstairs to help him settle into the capsule and seal the canopy door. Sturckow's audio and video feeds were piped into the ready room; we saw and heard what he saw and heard. He switched on the iPad that he and the pilots used to log flight data. Someone else's Solitaire game was still open. Stucky had used the tablet last. "You had time to play Solitaire?" Sturckow asked.

Sturckow did a run. Afterward, the tech asked Sturckow over the radio whether he had any discomfort, pain, heart issues, breathing difficulties, visual effects, or lingering dizziness. (Stucky regarded the centrifuge as an inexact though useful training tool, because "it twists your brain to make the g-profile work.") Sturckow reported some minor "grayout" on descent, but nothing of concern.

Grayout was the mildest manifestation of g-induced stress. Virgin expected passengers could experience some grayout. "And we may, once in a blue moon, have someone who's on the edge of having more than grayout"—ranging from tunnel vision, to brief G-LOC, or loss of consciousness, said Beth Moses. It would be up to her, as the chief astronaut trainer, to help customers avoid passing out. "We're going to show them how to keep some blood in the back of their eyeballs," she said. Each passenger would get a towelette, just in case.

Sturckow returned to the ready room, where Masucci was reclining on a couch.

"I've never had full-on G-LOC," said Masucci.

"Only once," said Stucky. It happened on his first solo flight, in 1981: he strapped into a T-34 prop plane, climbed to 9,000 feet, flipped the airplane upside down so the nose was pointing at the ground, and was intending to complete the maneuver, a kind of vertical U-turn, when things went awry. "Here comes the tunnel vision, marching in really slow. All I'm looking at is the g-meter. Then that disappears. Then I lost all vision. Before I know it, I'm like this"—Stucky pretended to be asleep in his chair, then jolted awake. He still didn't know how long he had been unconscious, but he was lucky he snapped out of it.

Stucky looked around when he finished the story.

Masucci, Sturckow, and Castleberry were staring at him with amusement and disbelief.

Stucky admitted that he didn't normally talk about that episode, for fear that a boss would worry about him falling asleep at the wheel. But enough time had passed, he said: "I figured you won't kick me out now."

I WAS UP NEXT. Castleberry asked if I was ready. Of course, I said, unconvincingly.

Sturckow told me not to worry if I felt sick. "It's no big deal. I'm not going to call your old man and say, 'Your kid's a wimp,'" he said.

The tech asked whether to set the profile for pilot or passenger. Castleberry suggested that I start at 50 percent passenger.

"I think he should do a hundred percent passenger," said Stucky.

"He did like seven g's in the Extra," said Sturckow. "He's good for it."

So Castleberry helped me into a spandex shirt with a biometric reader

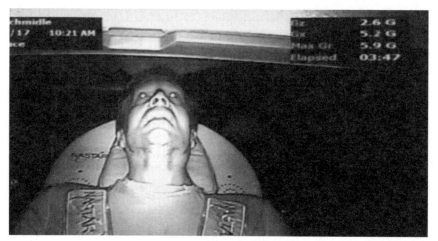

The author in the centrifuge, in 2017.

on the chest. She reminded me about the walnut and how to properly breathe, taking even sips of air.

"Just relax," she said, as she sealed the door behind her. An airtight silence.

I heard the tech commence countdown—"Three, two, one: FIRE!"— and then felt an immediate and tremendous pressure on my chest and ribs that pinned me to my seat. My heart rate jumped while the skin on my face bunched like bedsheets near my ears. But then I felt a sudden and unexpected wave of euphoria, the kind you hear marathoners describe when they are twenty miles into a race and should be feeling pain.

I completed the profile. I could have unstrapped, but I was curious enough to do it again. Stucky suggested I try the pilot setting, to feel the additional g-z on descent. I agreed, because nobody wants to be a wimp. But the pilot setting was considerably less pleasant, and unaccompanied by any euphoria. My vision telescoped and dimmed but didn't quite go out. Enough for me for one day.

Stucky took a turn next. The techs reconfigured the computer in the capsule to emulate an F-15, so Stucky could practice entering and recovering from a "departure"—a loss of control. The centrifuge spun to life and once it got going Stucky, using the flight controls to push the nose over into a dive, applied full left rudder, idled the left engine, and put on the right afterburner. He started spinning and tumbling with dizzying velocity.

"Holy shit!" said Leland.

"He's got a pretty good gyro rate going," said Sturckow.

But after ten seconds of wild spinning, the centrifuge froze. Stucky was left upside down, hanging by his shoulder straps. Leland explained that when the computer sensed that a pilot was about to injure himself, or break the machine, it automatically stopped.

Stucky made the centrifuge go TILT.

LATER, WHEN THE pilots were packing up to leave, a news alert popped up on Stucky's phone. He read it aloud to Masucci and Sturckow, a short item recapping comments made by Jeff Bezos at a satellite conference in Washington, DC, about Blue Origin's latest rockets.

A public appearance by Bezos, the world's richest man, was almost news itself. Bezos didn't share Branson's appetite for publicity. "Slow is smooth and smooth is fast," Bezos liked to say, echoing the military aphorism. If he was out and talking about rockets it probably meant that they were close, and if they were close it meant that Virgin's presumptive lead in the suborbital space race was in jeopardy.

Blue Origin had made significant progress in recent years. They had by now completed five successful, uncrewed spaceflights. And it wasn't just that they did it, but *how* they did it: their New Shepard rocket, like the ones at SpaceX, not only launched vertically but *landed* vertically. One minute the rocket was climbing straight up, the next it was speeding toward the Earth, only, at the last instant, to reignite its engine and hover momentarily before easing down. Imagine dropping a pencil, and it landing, erect and upright, on its own eraser.

On launch days, Bezos wore a pair of lucky cowboy boots with the words GRADATIM FEROCITER emblazoned on the side. It meant, in Latin, "Step by Step, Ferociously." In spaceflight, said Bezos, "You can't cut any corners." But, he added, "we don't have forever." Bezos was focused, urgent, meticulous, and had no shortage of resources. According to one calculation, he made an average of $215 million a day.

"I just keep waiting for them to announce that they flew humans to space," said Sturckow.

STUCKY WAS ABOUT to lose another race. This one was personal.

In 2009 Stucky and an Air Force lieutenant colonel named Jack Fischer made a wager over who would get to space first. Stucky had just joined Scaled, and Fischer had just joined NASA. The victor would win an evening of margaritas and tacos at Domingo's, a Mexican restaurant near Edwards Air Force Base. It was a good bet: the battle of publicly versus privately funded space travel, reduced to a duel between two old friends.

When they first made the bet it looked like Stucky would win. NASA was sclerotic, and Fischer knew it. It pained him to witness an agency that had once inspired the world become "plagued with dated paradigms, abysmal acquisition performance, a growing list of hazards," and an ever-shrinking budget. The thrill was gone—the Shuttle program was on the ropes; the US government was about to outsource its launch capability to Russia, a longtime foe; and Americans couldn't have cared less. It was a tragic tale of apathy eroding imagination.

Stucky may have won had SpaceShipTwo not crashed and set them back, but now it was up for grabs.

Two years after the crash Fischer emailed Stucky and said he was scheduled to launch in about a year on a Russian rocket to the International Space Station. It was July 2016; Stucky thought Virgin had a fair chance of being spaceborne by then, too.

"The flight/race/bet is ON," wrote Stucky. He and Fischer each promised to carry an item into space for the other person; Stucky provided Mike Alsbury's old name tag, which read MINI. (Alsbury was the younger of two pilots at Scaled named Mike.)

But the Russians had stayed on schedule, while delays added up and ate away at SpaceShipTwo's progress.

In March 2017, a week after Stucky got home from NASTAR, Fischer was on his way to Moscow. It burned Stucky up—how Virgin had squandered all that time, but also how America was reliant on Russia to get NASA astronauts into space—a sad sign of national decline. Part of Stucky was jealous of Fischer and wished he was the one about to squat into that old Soviet workhorse, the Soyuz. But another part of him cringed when he saw Fischer posting pictures of himself on social media from a press conference held in front of Lenin's statue, a prop for Vladimir Putin's PR machine.

Fischer abided by Russian traditions, and they had plenty; anything Yuri Gagarin had done, he did, too. During training, Fischer bunked at Star City, where he ate goulash to sustain him through long days learning Russian and the intricacies of the Soyuz. After Moscow, he flew fifteen hundred miles to the spaceport in Kazakhstan—an American marooned on the dreary steppe. How could his countrymen know that he was about to blast into space when he did so from a place so remote the *Washington Post* said it "muffled their launches to the point of obscurity for much of the American public"?

On April twentieth an orthodox priest blessed Fischer with holy water before Fischer boarded a bus, just as Gagarin had. The bus made the short trip to the pad, and on the way the bus stopped and Fischer got out, like Gagarin, and peed on the rear tire, like Gagarin. "It's an entire flow of tradition," said Fischer. He decided it was better to go with the flow than resist. "You could *not* do it, but do you really want to take that chance? Like, for real? I'll pee on that tire. Why not? Just in case it helps."

He climbed into the capsule and listened to a playlist he made. It included "Danger Zone" and "Fly Me to the Moon," his wife's favorite song. He hung a talisman, another Russian tradition: a felt sun with multicolored rays, the logo of the Houston hospital where his teenage daughter had been treated for a rare form of thyroid cancer. When the rocket ignited, the capsule rumbled to life, like "a sleeping dragon waking up in a bad mood," as another astronaut put it. The Soyuz boosted Fischer into space, and six hours later it docked at the ISS.

Stucky sent me a text message that afternoon. It was all over, he said: Fischer won the bet. And while Stucky hated to lose he was happy for his friend and pleased to know that now there was a memorial to Alsbury floating in space.

"Mini's name tag and photo are now in orbit," he said.

23.

ARTIFACTS

SIX MONTHS PASSED. Virgin still wasn't in space, and Richard Branson was getting restless.

In October 2017 he appeared at a conference in Helsinki. He felt fit and ready to go; he had rehabbed an old shoulder injury (a kite-surfing accident) and was playing a lot of tennis. "We are hopefully about three months before we are in space, maybe six months before I'm in space," Branson said at the conference.

His host nodded, and the statement triggered a pop-up alert on my phone. But it was another empty promise, divorced from reality; thousands of miles away, in Mojave, a maintenance technician had just shot compressed air into the wing cavity to blow out some drill shavings and accidentally popped the bond on a carbon-fiber panel. The repair was going to take weeks.

Branson knew that people snickered about him and his promises. "It would be embarrassing if someone went back over the last thirteen years and wrote down all my quotes about when I thought we would be in space," he said. But he was uniquely unfazed by embarrassment. He wore lipstick for one promotional campaign, and a wedding dress for another. He stuttered when he spoke without notes. He shared unflattering details about his sex life, how he and his first wife had "a bizarre sexual allergy" to each other. "Whenever we made love a painful rash spread across me which would take about three weeks to heal," he said. "We went to a number of doctors but we never resolved the problem. I even had a circumcision to try to stop the reaction."

It somehow all added to his charm.

Branson's detractors saw his loose statements as freewheeling and insincere, but he saw them as tools he could use to hold people's attention: so long as journalists were writing about him and his latest wacky statement, they weren't writing about his competitors; and so long as his employees knew that he expected to fly yesterday, they had an added incentive to hurry up and deliver.

"If you are an optimist and you talk ahead of yourself, then everybody around you has got to catch up to try and get there," he said.

That made sense or at least it sounded good—and maybe they taught this at business school in lectures about group psychology, about how to influence friends and make money. But Stucky wasn't sure that it actually worked. Or if it did, that it was actually trickling down.

Back in April, shortly after Stucky got back from NASTAR, the Space-ShipTwo program engineer sent out an email about how his team was working hard but didn't expect to meet any of their upcoming schedule goals.

It sounded like an admission of defeat. Stucky lost his mind. "We have reached the point where we must stop de-scoping timelines because we are overworked and understaffed and instead should be jumping up and down on senior management's desks saying exactly what contractor support, new hires, redistribution of effort, or whatever else is required to meet our tasks on schedule," he wrote. "We must stop hoping for the best and playing schedule chicken in the hopes that it will be somebody else's fault for the schedule delay and thus we'll be blessed with more time to complete our individual tasks."

He didn't want to be the guy berating his colleagues and accusing them of being lazy and unmotivated, trying to lead people who didn't want to be led, but sometimes it had to be done. At least he had an admirer in terse Sturckow, who commended Stucky for "an excellent fucking job" keeping "this team of innocents from wandering in the woods endlessly."

STUCKY MADE PLANS to meet Michelle Saling the week that Branson was in Helsinki. Stucky wanted to give her a photograph that Jack Fischer took from orbit, looking down on Earth with Alsbury's name tag framed in the porthole.

Stucky, Agin, and Saling met at Denny's in Mojave. Saling, a petite woman with straight brown hair, was wearing Alsbury's wedding band on a chain around her neck. Though she was tough, the last three years had nearly destroyed her. "No matter how strong I try to be, and how hopeful I try to be for my children, our lives are not complete without him," she said.

Saling had considered moving away after the accident, to escape the memories and reminders of her loss. But doing so would have entailed more disruption for her kids, and that's who she had to think about most now. Their friends were here.

Still, it was hard living in a small town twenty miles west of Mojave. Jets were constantly roaring overheard, triggering something in Saling and in her children each time. One day her daughter, Ainsley, came home in tears after four military jets streaked the sky at midday to commemorate the seventieth anniversary of Chuck Yeager's maiden supersonic flight, a moving gesture but one that reminded Ainsley of her dad. Another day her son, Liam, got assigned to write a paper about commercial space travel. "He came home and said, 'I don't want to do this,' so I told his teachers, 'I will not make him do this,'" Saling recalled. "Some things are just too raw."

Saling slid into the booth across from Stucky and Agin. She made a confession: she had driven through Mojave a few times at night, but this was her first time in daylight since the crash. Why would she? There was nothing for her here. She was angry with how Virgin treated Scaled and her husband, prioritizing its corporate survival over taking any responsibility. (Virgin was happy to share credit for Scaled's success, but in its submission to the NTSB, Virgin wrote, "Scaled Composites was responsible for all aspects of the flight test program at the time of the accident." The NTSB concluded that the crash was the result of human error.) And while Branson and Whitesides would talk about "sacrifice," they rarely mentioned Alsbury by name, like they wanted to put that chapter behind them, pretend it didn't happen, which left Saling in an awkward spot with her children as she wrestled with whether it was better to protect them or inspire them. Was she being fair discouraging Liam from pursuing his dream of becoming a pilot? Should she have left to go somewhere else and start a new life?

Saling had never been to the crash site. Stucky offered to take her, though first proposed a pact: they always cried when they saw each other, he said, and it was inevitable that it would happen when they got to the site, but, "This is a healing time. Let's not spiral and make ourselves hysterical." Stucky said this as much for himself, as he did for her. After he paid the check, the three of them piled into Stucky's SUV, and drove out into the desert.

Every time he went out there he seemed to find something new, and he was reminded of a grisly story he heard while he was in Iraq. The story came from a friend whose uncle had been a test pilot and astronaut, and who lost two friends in the eighties when their F-4 crashed in the English countryside. The uncle had been at the site, which reeked of kerosene and where the authorities were gathering human remains in plastic bags, and, months later, the uncle went back to polish a brass plaque that had been nailed into the side of an old oak tree to commemorate the fallen pilots. While he was standing at the tree, a dog ran up with a human hand in its mouth. Stucky couldn't shake the image.

He parked in the dirt on the side of the road. He and Agin and Saling split up and wandered silently into the desert. A recent storm had shifted the top layer of earth, exposing artifacts from the crash: a bundle of wire here, shards of cockpit glass there.

Saling picked up a piece of the wreckage, clutched it to her chest, and began to cry. Stucky put his arm around her. She was obviously disturbed, and Stucky offered to drive her back to Mojave. But she declined and said she wanted to go see the other impact points.

In the car, on the way back to Denny's, he told her, "I know you don't want to hear that something good came of all this." But if they learned anything from the crash it was the importance of crew resource management, of scripting every word and action during the flight. He had recently gone back to NASTAR with the other pilots and they had all committed a blunder, calling out "trimming" when they were still subsonic. He explained to Saling that if they trimmed—entering the gamma turn, going sharply nose up—while subsonic, before reaching Mach 1, it would aggravate the transonic pitch-up and could lead to a crash. So when Stucky got back from Philadelphia he made the engineers reprogram the altitude and airspeed dials to turn blue when the spaceship went supersonic; the color prompt would indicate that it was safe to begin trimming.

"Maybe Mike's death saved eight people from dying on a later flight," Stucky told her. "I'm not telling you that to try and make you feel better. Just that we've made some serious changes. Because ninety-nine percent isn't good enough."

She thanked him, because for all of her struggles with the loss, she was also struggling with the why. What was her husband thinking? He was a great pilot. Why did he do what he did? He knew better.

She therefore found it oddly comforting to hear Stucky recite examples of others' mistakes, revelations of others' fallibility, to know that it could happen to anyone.

24.

FULL DURATION

THE ROCKET MOTOR WAS READY. That was one of the cruel ironies of the crash, how after years of being hamstrung and delayed by propulsion problems, now Virgin had a suitable rocket motor but no spaceship to put it on.

I once went and watched a "hot fire" with Moses. A hot fire is a live test of a rocket fixed to a ground stand. It was December 2014, five weeks after the crash. We drove out to an area north of the airport, beyond the blacktop where airlines ditched old passenger jets—a graveyard of fuselages, out among the dirt mounds that looked like giant anthills where local companies buried sensitive chemicals.

Moses had warned me beforehand that they were testing a rubber-fuel design, one Virgin had developed internally, and completely different from Luke Colby's plastic-fuel design, but that it was only a test, so anything could happen.

Despite Moses's warning, the rocket lit beautifully and shot a plume of fire forty feet into the air.

Moses was encouraged, though the propulsion team still needed to tweak the flow rates to ensure that the motor would burn strong and steady for sixty seconds, all the way to the end—all the way to what they called "full duration." The final seconds were the crucial ones, said Jarret Morton, who had earlier advocated for firing SpaceDev and was now

Virgin's lead propulsion engineer. "You're going the fastest, and you're the highest, and you're out of the atmosphere, and you have the least drag, and your weight is the lowest," he said. A single sputtering second could mean the difference between reaching space and fulfilling a dream—or falling short.

Over time, Stucky got increasingly involved with the propulsion effort. When Morton decided that Virgin's test stands—a horizontal one and the vertical one I saw with Moses—were inadequate because they didn't accurately represent SpaceShipTwo's flight path—the ship flew neither level to the ground nor with its tail pointing up—Stucky brought Morton to China Lake, which had a rotating test stand, to meet senior military officials on the base. Stucky and Morton inquired about using the rotating stand. The officials said they wished they could help but couldn't afford to run the risk of Virgin accidentally blowing up the stand and leaving the Navy with no way of testing its submarine-launched thermonuclear ballistic missiles.

This left Morton with no choice but to dig a hole. His team burrowed an L-shaped pit about 14 feet wide and 60 feet deep, with a dogleg at the bottom to redirect the flame out of a duct 140 feet away. In May 2017, they finished digging and were ready to run their first test.

Stucky drove out past the old jets and the mounds and parked behind a berm with a pair of binoculars. Dillon joined him. His son had become a proficient drone pilot, and Stucky had helped him get authorization from the FAA to operate in the airspace over Mojave. Dillon parlayed this into contract work with Virgin. He was good and was soon filming commercials for car companies. Officially, Stucky was Dillon's "escort" on the site, though he regarded the experience as another opportunity to bond with his son.

Dave Raibeck, a propulsion engineer, crouched with them behind the berm. Morton was monitoring the test with his team from a bunker near the stand.

The rocket motor roared to life. Stucky counted off the seconds—
Thirty . . . forty . . .
He admired the steadiness and density of the flame as it shot out of

the duct because it suddenly felt very real, that this was the rocket that was going to blaze him into space, that they were going to do this after all—

Fifty . . . sixty . . .

And when the flame snuffed out, Stucky whooped, "Yeah! Baby!!" hugging Raibeck and Dillon, while Morton emailed Moses and Whitesides with the results that he knew they were hoping to hear.

"Test complete," Morton reported. "FULL FUCKING DURATION."

BY AUTUMN 2017, *duration* had become the word on everyone's lips: How long should they burn the rocket on SpaceShipTwo's first powered flight? Typically, these debates started with hard, objective numbers, but they often devolved into subjective discussions about risk.

Stucky was prepared to open the throttle and burn the motor for the full sixty seconds. But knowing that he'd find few backers, he lobbied for forty. He felt his negotiating position was strong: he was going to be the one in the cockpit after all.

Stucky created slideshows and animated presentations, arguing that anything less than forty seconds would put him and the copilot at unnecessary risk; a shorter burn, incidentally, would condense the pilot's "workload" and also keep the ship in that unstable transonic zone longer than desired. Stucky wanted to get in and get out of there as quickly as possible. A longer burn would allow just that.

His recommendation met opposition. Just because the motor *could* burn that long didn't mean it *should* burn that long. One engineer expressed concerns about overheating and proposed twenty-two seconds. Another was worried about transonic flutter and recommended fifteen: "I don't feel we are being restrictive in the reach to space, just being sure we don't take monster steps to miss a potential disaster."

Stucky took offense. "Your shotgun e-mail makes it look like I'm trying to run a 'Damn the torpedoes, full speed ahead' approach to flight testing," he replied. He was always ready with some aphorism or arcane nugget of flight-test trivia to make a point. He went on, "To paraphrase Harrison Storms, the North American Aviation project manager for the X-15 as well as Apollo, we need to work with thoughtful courage and not be blinded by fearful safety."

He assumed he could count upon support from the other pilots but found dissent among the ranks. Ericson, with his rosy cheeks and eye for posterity, thought forty seconds was too long and too risky for the first test. He and Stucky had squabbled over a number of recent issues, though, to be fair, Stucky squabbled with others, too. "He pisses people off," said Moses. Sturckow would complain that every time Stucky took the controls it became all about him, turning into what he called "the Forger Shit Show."

Stucky brushed the enmity off. It was natural. If you took any handful of accomplished pilots with sizable egos and put them in an environment where opportunities were scarce, you had to expect some backbiting. "If you're in the squadron that's only getting a few hours of flight time a month, morale is going to suck in the ready room," he said.

But the enmity between Stucky and Ericson was real. Stucky saw Ericson adopting the same hand-wringing tendencies he'd previously warned against, and Stucky didn't trust him.

The feeling was mutual: Ericson thought Stucky was being reckless and only looking out for himself.

Burn time was Virgin's "most important programmatic decision," said Ericson, and they couldn't afford to be flippant. He noted that SpaceShipTwo was expected to experience some "lateral directional instability"—aggressive yawing—as it approached Mach 2, which was, as he put it, "an unforgiving environment." Moreover, Ericson protested, he was the one who had to sign his name on the paperwork that went to the FAA, verifying that they were ready: it was his neck on the line.

Stucky found Ericson's contention laughable. A signature on some paperwork? Stucky was going to be the one in cockpit! "That's like the hen telling the pig, 'I have to lay the egg in the bacon-and-egg sandwich,'" he said. "I might not be signing my name for the FAA, but I'm carrying a lot more risk than his ass."

Their feud eventually boiled over when Ericson found out that Stucky was essentially ignoring him and continuing to rehearse in preparation for a long burn. Ericson got up in Stucky's face, crimson with rage. "That's a foul!" said Ericson.

Moses had to step in and mediate. He reminded Stucky and Ericson that they were *almost* there, that once they shook out "the last few remaining gremlins" and got their final glide flight off, they could resume powered flight tests—and be ready to take aim for space.

In late 2017, Moses wrote to his team, "I often say that rockets don't get us to space—people do. We have a great group of people, so space should be right around the corner. . . . I think you all can feel it, like I do, that the thing we all came here for—flying people to space from Spaceport America—is within reach."

AMID THESE ONGOING and testy debates, Stucky got invited to Idaho. Burt Rutan, the sage Scaled founder, had retired there and bought a majestic cabin on a pine-forested bank of Lake Coeur d'Alene. In a nearby hangar, he was trying to finish what he insisted would be his final project—a high-wing-mounted prop plane that could land on ground, water, and snow. He called it the SkiGull.

Rutan had been working on the SkiGull for four years. Some days had been promising. Other days he wanted to savage the plane with a hammer. Now he was nearly done, and he needed someone to flight-test the plane. And not just anyone. "It takes a Real Test Pilot to do the research flight tests needed to develop a complex addition to the flight control system," he wrote to Stucky. "Hope you can make it."

Stucky was flattered. He didn't have a close personal relationship with Rutan, so he read the invitation as a testament to his skills. But the timing could not have been worse. It was considered bad form to throw yourself heedlessly in harm's way before a mission, like the astronaut who crashed and died doing stunts in the antique plane six months before his scheduled Space Shuttle launch. Stucky thanked Rutan for the opportunity but replied, "I've got some cool SS2 fights coming up and would rather not hurt myself or die testing skigull!"

But back in FAITH the engineers had no sooner sealed up the wing that had been damaged by the compressed air mishap, when the horizontal stabilizers suddenly stopped responding to the controls. This left everyone puzzled and was going to take time to figure out.

Stucky felt suddenly torn between his philosophy of living life to its fullest and his professional duties at Virgin. He was reminded of something a NASA psychologist told him during one of his screening interviews, when she had wished him good luck and said, "I hope you get it. I hope it's everything you want it to be." Her words sat funny with him at first, but the more he thought about it, the more he realized she was saying that he shouldn't make space flight, as he put it, "the 'Be all, End all.'" In other words, he shouldn't avoid doing the other things he loved.

"I've been on this program pushing ten years now, and another decade I had chased being an astronaut. If I had stopped doing the things I enjoyed doing decades ago, I would not have had as rich of a life," he said. He and Dillon had recently gone paragliding in the French Alps, soaring in the shadow of Mont Blanc by day and eating baguettes and cheese and wine in their lakeside camp at night. "I want to be able to spend time in the mountains with my son," he said.

A month after initially declining Rutan's offer, Stucky and Agin flew to Idaho. She loved the Northwest. They had traveled to Oregon to watch the August 2017 solar eclipse, just him and her with their camp chairs and eclipse shades on a remote gravel path that he had identified as being a prime spot based on all the meteorological data he could scour. Now, pulling into Coeur d'Alene unannounced, Stucky contacted Rutan and said he was in town and maybe he could take the SkiGull for a whirl.

Stucky flew it the next day. It was choppy at first, but he got the airplane under control, and Rutan was delighted.

Rutan suggested they meet for breakfast the next morning at the resort where Stucky and Agin were staying. They took a table overlooking the lake. Rutan asked how things were going back in Mojave. Stucky described the ongoing debate about how long to burn the motor.

"You've got an airplane that should be able to run a full burn," said Rutan. "Plan for a full burn and let the pilot abort it whenever he can. The

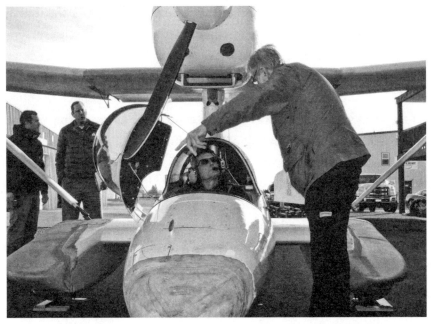

Burt Rutan (right) with Mark Stucky in the cockpit of the SkiGull, in 2017.

reason we have a hybrid with an off-switch is so we can turn it off any time we want to."

Besides, said Rutan, "You'll enjoy that next minute of burn." Chuck Yeager once described the gleeful sensation of riding a rocket through the upper atmosphere and the temptation to keep going: "You're bug-eyed, thrilled to your toes, and the fighter jock takes over from the cautious test pilot. Screw it! You're up there in the dark part of the sky in the most fabulous flying machine ever built, and you're just not ready to go home."

"It might be the only chance you have to be an astronaut," said Rutan. "What are they going to do? Fire you?"

"Probably," said Stucky.

"Fuck no! You'd be a hero," said Rutan.

"What are they going to fire you for?" Agin asked, interrupting her conversation with Rutan's wife, Tonya.

"He's trying to tell me to just go for it on the first flight," said Stucky.

"If he's already done the real dangerous stuff"—passing through the transonic zone—"and everything's smooth, just run it out," said Rutan.

Just one thing, he asked: "Promise that you will never unlock the feather during boost."

Deal, said Stucky.

A FEW WEEKS later Stucky was 48,000 feet up in the air, preparing to dive. It was January 11, 2018. The engineers had finally fixed what Moses had to come to despairingly call the "confounding" h-stab problem, and Stucky was about to conduct the final glide flight.

This was not a modest endeavor. The engineers wanted to stress the vehicle as much as possible prior to their first supersonic flight; the best way to do that was a steep dive, like the one Stucky entered in 2011 that put him into the inverted spin. Everyone thought the powered tests were hard, but Stucky found the high-speed glides far more worrisome.

Cleared for release, Stucky came off the hooks and pushed the stick forward. The ship went into a weightless dive. Stucky floated in his shoulder straps.

As the arm of the Machmeter spun clockwise toward, but not quite reaching, Mach 1, they crept into the transonic zone. Shock waves buffeted the sides of the ship. Stucky imagined being trapped inside a drum as it was struck by hundreds of rubber mallets. He calmly proceeded through the test card.

"Roll rap," he called out—banging his palm against the side of the stick, to highlight any roll-related structural concerns.

"Pitch rap," he said—slamming his palm against the butt of the stick, to do the same for any pitch-related concerns.

He felt the spaceship wobble.

"Catching it," he said to his copilot, Mike Masucci.

The buffeting continued, and now, with SpaceShipTwo approaching 22,000 feet, Stucky tried to pull out of the dive. But when he did he got a flashing red light, indicating an h-stab malfunction: the left h-stab was stuck, with a five-degree split between the angles of the left and right stabilizers. The ship wanted to roll upside down.

Stucky knew he had to act, and fast, but just then a flash of sunlight filled the cockpit. It briefly blinded him.

"I can't see! I can't see!" he said.

Fortunately he knew to switch to backup power, which allowed the left h-stab to unstall. They glided safely back to Mojave.

Objectively, the flight had been a success. "We got the kick-ass hard test points done," said Stucky. But that h-stab worried him; if it got stuck when he was under rocket power, going Mach 2, it might just send him barrel-rolling into danger.

Afterward, Whitesides hosted a small ceremony on the runway. But Stucky didn't feel much like celebrating. His heart was still thumping; he knew he had just cheated death again.

And now they were about to start the really dangerous stuff.

PART THREE

SONS

25.

SLIPPING THE SURLY BONDS

STUCKY MADE HIS OWN PLAYLIST. He knew it was cheesy but he didn't care: he blasted Elton John's "Rocket Man" on his way into work. The predawn sky was full of stars.

He pulled into town on the morning of April 6, 2018, past the roach motels with flickering signs. He turned into the airport and saw the security guard up ahead holding a clipboard. Stucky lowered the volume on the music and rolled down his window.

He provided his name to the guard.

The guard checked his list. Nothing.

Stucky asked him to look again.

Nothing.

Eventually, one of Stucky's colleagues ran over and told the guard, "You have to let him in. He's the pilot."

The guard relented and waved Stucky through.

STUCKY PARKED AND walked across the lot, into FAITH's large beige hangar. He and Mackay, the graybeard, flew the simulator once, then walked upstairs to the mission control room, where three dozen engineers sat at their consoles, reviewing the latest flight data. At their supervisor's direction, a representative from each engineering discipline confirmed that they were ready to fly the day's mission: a thirty-second burn, to about 80,000 feet.

"Flutter: Go."

"Aero: Go."

"Stabs: Go."

"Pneumo: Go."

"Avionics: Go."

"Prop: Go."

"Thermal: Go."

"Loads: Go."

Stucky knew that some of the engineers were nervous. But he also knew that they thought he carried a bus under each arm, and he wanted to break the ice, to remind them that they were in this together, that he and Mackay believed in them. So when they were done confirming that they were ready, Stucky turned solemn and addressed them, on behalf of himself and Mackay, saying, "We love you—no shit."

He and Mackay got on well. Mackay was only a year older than Stucky, but he possessed a patient wisdom beyond his years. They were like a comedy act, with Stucky playing the prankster to Mackay's earnest straight man. Their relationship had been like this from the beginning. When Mackay got hired and first came out to Mojave, Stucky was working at Scaled. They met, and Stucky asked Mackay if he knew what Mojave meant "in Indian." Mackay had no idea—he was from Scotland—so Stucky fibbed and told him it meant "Big Wind." Stucky didn't think of his innocent fabrication until years later, when he was sitting in on an interview with Mackay and a prospective hire. They were all talking about the weather and the wind, and Mackay remarked, confidently, that, after all, Mojave was Indian for "Big Wind." Stucky suddenly realized that Mackay had taken him at his word: "Big Wind" was becoming Mojave folklore. Stucky sprayed coffee from his nose.

Moses was also there, in the adjacent mission control observation room. He was proud of where they were and what they were about to do, how they had been written off but never quit. He wondered if the anticipation he felt was how rock and roll bands felt backstage as they were about to come out and play a stadium concert. "We're eager to get out there and show everybody how good we really are," he said.

Even Ericson had encouraging things to say. He told the team that it was normal to have "a little pit in your stomach," but to remember that they were "extremely well-prepared for this sortie." He went on, "My only words of advice are the ones my commander gave us before we flew for Kosovo, Night One. His words were, simply, 'Step on time.' Which meant, 'You guys have trained. You've planned. You've prepared. You've got this in the bag.' Stick to the plan and it's going to be a great day."

But nerves were inevitable. Stucky and Mackay were about to strap into what was essentially a winged bomb going more than eleven hundred white-knuckling miles an hour—nearly twice the speed of sound. Stucky had flown SpaceShipTwo's three successful rocket-powered missions, the fourth crashed and burned and killed his best friend, and now he was about to fly the fifth.

I shared a glass of prosecco with Cheryl Agin the previous afternoon. She had been out all day running errands and picking up Stucky's flight suit and getting her hair done to stay busy and distracted. She was ready to get this over.

"It's a little scary," she said. "Because of Mike, and because it's just been so long coming. I'll be glad when this one is behind us."

IN HIS HASTE to leave the house that morning, Stucky somehow forgot his flight card, with all of his notes and annotations. He left it on the kitchen counter, which he had realized while he was in the car, a little before five. "Any chance you can be here by 0645?" he texted Agin.

She hurried and made it. When Stucky left the mission control room and headed downstairs to retrieve his helmet bag from the locker room, Agin was waiting for him in the lobby with his flight card. She gave him a kiss and was escorted to a viewing area by the runway with the other families.

Stucky and Mackay were driven to the east end of the runway, where WhiteKnightTwo and SpaceShipTwo had been towed. The lime-green fire engines operated by men in hazmat suits moved into position. The sun rose over the buttes to the east, and a faint half-moon hung over the Tehachapi Mountains to the west. The wind turbines spun on the nearby hillsides.

Stucky climbed through a side hatch into the left seat of the cockpit. On Facebook, he posted a message: "Zero hour 9 am and I'm going to be high as a kite by then." He put on his helmet, lowered its shaded visor, slipped on his gloves, and breathed through his oxygen mask, to ensure that it was working.

At eight a.m., WhiteKnightTwo powered up its four jet engines and charged up the runway. It climbed on its massive 140-foot wingspan and flew north over Death Valley, banked west toward Sequoia National Park, and then south, back toward Mojave. Stucky looked out the window of SpaceShipTwo, wondering if Agin and his kids, who were watching from the runway, could see him through the spotty clouds.

The SpaceShipTwo cockpit.

He took a deep breath and extended his right hand. Mackay grabbed his hand and held on: their fates were now intertwined.

A radio operator gave them clearance to commence "L-4" checks: four minutes until rocket launch.

"Roll boost?" asked Mackay.

"On," said Stucky.

"Speed brake?"

"Enabled."

"Dampers?"

"On."

They were ready.

Two minutes. One minute. Thirty seconds.

Mackay flipped the switch that controlled the release mechanism. "SpaceShip is armed, with a yellow light," said Mackay.

Stucky gripped his fingers around the stick and heard the WhiteKnightTwo pilot counting down in his ear: "Three, two, one: release, release, release."

STUCKY WAITED HALF a second for WhiteKnightTwo to peel away before calling "Fire!"

Mackay lit the rocket. The thrust from the boost pushed them against the back of their seats, an exhilarating sensation that made Mackay think of riding a magic carpet.

They broke through the transonic zone, and ten seconds into the burn they were going faster than the speed of sound. Stucky began trimming the h-stabs for the J-shaped gamma turn; soon, the ship was flying nearly seventy degrees vertical. Sensors attached to Stucky's body showed his heart rate jump forty beats a minute during the first few seconds after the rocket fired, but he felt eerily calm, like time had become an elastic concept that was stretching to let him think—a phenomenon described by another test pilot as akin to "time dilation."

Soon he and Mackay were traveling at Mach 1.8, about twice as fast as a Tomahawk cruise missile. Outside the vehicle, the light was draining from the sky, turning it a deep, muddy blue.

But as the spaceship cleared 60,000 feet Stucky suddenly felt the wings tipping, like the ship wanted to flip upside down.

"A little rolly there," said Stucky.

This was an understatement. He was straining to keep the wings level and he knew that he should abort—manually shut down the rocket motor—before he lost control of the spaceship, because entering a spin or a tumble at supersonic speeds would be hard to recover from. But he was electrified by the ascent and believed that he could hold the spaceship steady, so he let the motor burn on, toward its planned duration of thirty seconds.

A moment later, Stucky felt the vehicle roll again, even harder this time.

"Ahhh!" he groaned. SpaceShipTwo was slipping beyond his control. He could feel it wanting to cork, and he finally called out, "Abort, abort, abort!"

Mackay leaned forward to hit the mushroom switch to abort when the rocket motor's onboard timer hit thirty seconds, automatically shutting down.

Stucky got the craft under control and looked out his window. The Earth's bright-blue surface filled his porthole. It was a stupendous sight: the outer edge of the atmosphere was dancing with wispy tendrils. The altimeter showed 84,000 feet—higher than Stucky had ever been—and he could now testify to the awesome power of the "overview effect."

But then he realized that he wasn't supposed to be looking down at Earth—that this was the plan for tourist flights, but for this test the ship was meant to stay upright.

He was momentarily confused: he should have been seeing only the blackness of space; somehow he had flipped and inverted without noticing.

"Oh, shit," he said. "The gyros are messed up"—his instruments, for some reason, still indicated that he was right side up.

"Totally," Mackay confirmed.

Stucky tried to reorient the ship by blasting thrusters of high-pressure air that they used to orient the vehicle in low-gravity environments. He then instructed Mackay to unlock and raise the feather. As the feather went up, the spaceship righted itself, just as it had on Stucky's harrowing glide flight seven years earlier: the innovation that led to the 2014 crash was, when properly deployed, a godsend.

Stucky searched for a stable guide on the horizon. "I got the moon," he said, as he glided down over the California desert, toward the constellation of dry lake beds below.

Dillon was down on the tarmac in Mojave, filming the flight. He powered up his drone and zipped it along the runway to intercept Space-ShipTwo as it made its final approach. The spaceship landed thirteen minutes after its release from WhiteKnightTwo. "That was about the coolest thing I've ever seen," said Dillon.

Michelle Saling was on her way to her daughter's school when she saw WhiteKnightTwo in the distant sky, its twin contrails like walrus tusks. She shed tears of joy for Stucky and sent him a message, "Mikey would be happy," but she also felt an upwelling of sadness: "Seeing that just intensified my sorrow that he's not here."

Jack Fischer, the NASA astronaut, posted a congratulatory tweet to "the whole @virgingalactic team," but to Stucky in particular who, he wrote, was "slipping the surly bonds like a boss!"—an allusion to the opening lines of "High Flight."

Oh! I have slipped the surly bonds of Earth.

When SpaceShipTwo landed, Stucky and Mackay climbed out of the cockpit, pumped their fists, and high-fived the crew. Stucky crossed the runway to where his colleagues were whistling and applauding. He hugged Agin.

Whitesides handed him a microphone.

"I don't know where to start," Stucky said. "Some of us have been waiting years to do that again!"

They were years late but on their way to being a vertically integrated spaceship company: flying rockets, with their own pilots, in their own spaceship, with their own rocket motor.

Dillon Stucky (left) taking a photo of Mark Stucky and Dave Mackay, in 2018.

"Richard has been waiting longer than all of us," said Stucky. "Hopefully, we gave him a good flight."

BRANSON AND HIS grandchildren had been huddled around an iPad at their mountain lodge in Verbier, Switzerland, watching a livestream of the flight. At the end of the rocket motor burn, the room had erupted with delight, when Branson's three-year-old grandson asked him why he was crying when everyone else was cheering. Branson recalled, "It was just such a beautiful moment."

They were finally conducting supersonic tests of their spaceship again. Branson knew that Virgin still faced plenty of challenges, that they would have to perform a few more envelope expansion flights before they shot for space, and even once they got to space they would have to repeat it several times to prove they were ready for commercial service. Still, Branson had been funding this project for fifteen years and felt that redemption was finally near. "We've proved skeptics wrong," he said.

He estimated having spent nearly $1 billion on the program, all of which he expected to recoup. "There are, we believe, millions of people who would love to go to space, and we want to tap into those people," he said. "If you can create the best—the best hotel chain, the best clubs, the best spaceship company—it'll become very valuable."

At that point, he valued Virgin Galactic and an uncrewed satellite-launching spin-off, Virgin Orbit, at "some billions of dollars," but said that they could soon be worth "*many* billions." He noted how the expertise and technology they were developing for suborbital space tourism could well extend to other industries. Perhaps Virgin rockets would soon be propelling passenger planes around the world in record time, he said. He was also planning to build a flying car.

STUCKY SAID WHAT he had to say, then passed the microphone to Mackay.

"I want to do that again!" Mackay declared.

He and Stucky could hardly believe their good fortune. Hundreds of customers were waiting to pay $250,000 to experience the overview effect once. Stucky, Mackay, and the other Virgin pilots could ultimately be bearing witness to it once a week. The most experienced astronauts had only completed seven space flights each.

They had a lot to review before the next flight. Like, Why did the gyros flip? But now it was time to celebrate. Stucky and Mackay went back to the hangar, where Mackay poured glasses of fine single malt.

Later, Whitesides and Moses invited the whole company into the hangar for an apple cider toast.

"Hell yeah! You did it. What a day!" said Whitesides.

Moses added, "We've been working for this for so long, and each one of you have put your blood, sweat, and tears into that thing that just went up to Mach one-point-eight. . . . You'll cherish that memory forever."

Afterward, Agin waited for Stucky on a couch in the lobby near the bowl with the bite-size chocolates. He eventually came down holding his tumbler of scotch, still wearing his flight suit. He flopped down on the couch beside her.

An employee and his son were taking photographs in the lobby; the son was posing behind a headless cardboard cutout next to a cardboard cutout of Branson in a spacesuit. The employee asked Stucky whether Stucky would pose for a picture with his son.

Of course, said Stucky.

Bill Kuhlemeier, the mission control room director, came down and sat on the couch. He and Stucky relived the flight, and Kuhlemeier asked about the view.

"We got the great shot when we were upside down," Stucky said, an image he had been dreaming of since he was three years old, watching John Glenn on live TV.

"That'll be worth blowing up and having on a wall somewhere," said Kuhlemeier.

Mackay came down in jeans and a T-shirt, and Kuhlemeier ribbed Mackay for not sharing his scotch.

"I ran out of glasses," said Mackay. "And Forger drank it all anyway."

"Thanks for telling me you're changing," said Stucky.

Then Ericson came down, jangling a set of keys. "Denny's?" he asked.

Everyone was famished. Denny's sounded great. As they were getting up to go, Mackay took a scrutinizing look at Stucky, still in his flight suit.

"You're not going like *that*, are you?" asked Mackay.

SpaceShipTwo.

26.

MILLION-DOLLAR VIEW

IKE A FOOTBALL TEAM reviewing game tape on Monday morning, Stucky, Mackay, Moses, and others spent the next few days analyzing the last flight. They got to the buttom of the gyro mystery: there had been a software glitch, or rather a patch meant to fix the glitch that Virgin's engineers had decided to ignore because the software developers deemed it noncritical and hadn't specified what the patch was meant to fix; the engineers hadn't wanted to risk fussing with a software update on the eve of a big flight. They thought they could deal with it after the flight.

The patch fixed the glitch.

It could have been a costly oversight, though Stucky didn't fault the engineers for it. He probably would have done the same thing and told himself that he was avoiding some unnecessary risk, like a marathoner who avoids wearing new shoes for the first time on race day.

Other, more worrying, problems demanded Stucky's attention. He and the rest of the team tried to figure out why SpaceShipTwo had gone squirrelly as it approached Mach 2. There were fears about this band of airspeed becoming as beguiling to them as the transonic zone had once been. The engineers were already working on a digital h-stab system that would automatically detect and correct undesired yawing or rolling, but that system was months from being completed, and they needed more data.

They looked for a short-term fix, and the data they got back from the first flight would certainly help.

They had gone into that flight blind. "You're always pushing against what you know, versus what you don't know, and how much of what you don't know matters," said Moses. "We had zero data."

But they theorized that climbing at a steeper pitch angle might alleviate some of the instability. Virgin made several modifications to SpaceShipTwo, like off-loading some ballast in the nose to shift the center of gravity aft and changing the "scarf" angle on the rocket motor nozzle. These mods, they hoped, would enable them to more aggressively enter the gamma turn.

It worked.

In late May, Stucky and Mackay flew SpaceShipTwo again, this time with Mackay at the controls. He pitched into the gamma turn at almost ninety degrees vertical, twenty degrees more than Stucky was on the previous flight. They burned the motor for the same amount of time but went 30,000 feet higher, reaching an altitude of 114,000 feet.

"Hang on!" Mackay said, as they roared into the darkening atmosphere.

THE ALTITUDE DISCREPANCY became a source of puerile amusement. Stucky got an email the next day with an attachment comparing snapshots of SpaceShipTwo, mid-flight, under his command, next to one under Mackay's command, with a caption that read: FORGER CAN'T GET IT UP.

It captured the jocular mood inside the company; two successful flights in as many months had lifted spirits. They had something to show for their efforts, apart from lofty promises and publicity stunts—a good ship and a great rocket motor and a talented team of techs and engineers and pilots. Years of labor and sacrifice were finally paying off.

They even had a durable philosophy in place. As Ericson liked to say, "Flight test is a process of predicting, discovering, and then making things better," which is what they were doing now.

Fly, test, notate, adjust.

On their next flight, they were aiming to go higher yet. Stucky planned to sit this one out. He and Mackay had briefly considered flying together again, but Mackay worried about the program becoming overly dependent on them. He knew the importance of getting others some experience in the cockpit.

Mike Masucci, the loping former U-2 pilot, was the first one up. He and Mackay prepared to burn the motor for forty-two seconds. This duration, the engineers calculated, would propel the spaceship to around 170,000 feet—into the thin, black air, well above the ozone layer, beyond anywhere Virgin had gone before.

Up there, the pilots would have decreased aerodynamic control. They would be more reliant on the reaction control system, or RCS, which consisted of a dozen nozzles along the outside of the ship that fired blasts of compressed air on command. Every spaceborne craft has had a version of RCS; otherwise they would drift, and for those in deeper space, add to the infinite count of heavenly bodies.

The company knew this would be a major envelope expansion flight, but they didn't know how much of what they didn't know would matter. Still, they assumed the sights would be spectacular, and word came down from upper management urging Mackay and Masucci to think of something profound to say from up in the mesosphere, something for the GoPro cameras that the team back in London could use for marketing.

Stucky considered this an absurd distraction, and his fears were confirmed when Masucci joked that he was worried about forgetting his lines.

Also, Stucky had concerns about how Mackay used the RCS; he thought Mackay's rapid-burst technique made it difficult for the copilot and those in mission control to distinguish intentional blasts from inadvertent blasts. It was important to have others monitoring your RCS supply. If you drained your RCS in the mesosphere, that would be bad, like running out of gas in the middle of nowhere.

Stucky conveyed his concern to Mackay. Mackay told him not to worry. Stucky let it go.

The flight took place on July 26, 2018. Stucky watched it from mission control. The last time he had watched a supersonic SpaceShipTwo flight from mission control it ended in catastrophe.

The mission got off to a strong start. Mackay said, "Fire!" and Masucci lit the rocket. Masucci kept an eye on the Machmeter and called out "One-point-five Mach" so that Mackay knew when they were coming up on Mach 2, and they were zipping along, nose pointed up at the sky. Mackay felt the wings tilt slightly, but he stuck with it, and on they climbed.

At the end of the forty-two-second burn, as they crested the top of their flight path, Masucci deployed the feather and took a moment to look outside. "Beautiful black sky," he said, as the spaceship floated through the mesosphere. "Million-dollar view out the window, Dave."

It was beautiful, indeed—a vision of Earth framed against the inky depths of space. But as mountaineers knew well, it was often just at such moments of placidity when terror struck. In *Moby-Dick*, Ishmael warns about the "subtleness of the sea," how a serene ocean surface disguised so much—"how its most dreaded creatures glide under water . . . treacherously hidden beneath the loveliest tints of azure."

The dreaded creatures of the upper atmosphere were closing in.

Stucky looked up at the telemetry monitors and saw that Mackay was firing rapid bursts of RCS, which was pitching SpaceShipTwo into a backflip.

And now it was beginning to roll.

They were quickly losing control and dancing in midair like a fishing bob on the top of a lake. Stucky feared that if Mackay couldn't slow their rates soon, their off-kilter descent could damage the ship.

"Working on this," said Mackay. The feather was already up, but the atmosphere was too thin for the feather to bite. The spaceship spun and tumbled.

Three years before Neil Armstrong landed on the moon he and David Scott, another astronaut, had faced a similar problem when their Gemini capsule started spinning in orbit because one of the RCS thrusters got stuck "on."

"We're tumbling end over end up here," Scott had said.

"Can't turn anything off," added Armstrong.

They were tumbling at 360 degrees per second, and had Armstrong not deftly recovered it was conceivable that they might have spun off into oblivion.

Stucky wasn't worried about that because he knew gravity would pull SpaceShipTwo back to Earth. Still, Mackay was running low on options and he needed RCS to right the ship, but he was fatefully low on that, too.

"Running out of thrust here," he said.

Stucky got on the radio and suggested they switch to backup RCS, which they did. But Mackay was quickly exhausting that supply, too. They were entering perilous territory. The spaceship's recommended pitch rate was no more than five degrees per second, but it was pitching at about fifty degrees per second. Finally, by ingenious design, the feather—the angel's wings and the source of the crash—clawed enough air to tip the ship upright, enabling a calm, feathered descent.

AFTERWARD, STUCKY WAS annoyed because, as he feared, Masucci had seemed more worried about remembering his lines than concentrating on the instrument panel. Had Masucci been focused on the panel he might have noticed the increasing pitch rate earlier and helped Mackay stop it before it got as bad as it did.

They had, like Armstrong, avoided catastrophe. (Armstrong, who kept remarkably cool at the moment of greatest despair, ultimately activated the secondary thrusters, which stabilized the ship, saving his and Scott's lives. "The guy was brilliant," said Scott.) But Stucky couldn't help thinking that Mackay and Masucci had recovered because they were lucky rather than because they were good.

The next day, Stucky stopped Mackay in the hangar and asked if they could speak privately. Stucky valued his relationship with Mackay, but he knew from his experience with Siebold how a festering animosity could undermine the broader mission. He told Mackay how much he admired him and his leadership and respected him in the cockpit, but he reminded Mackay that he had raised concern about Mackay's RCS technique before the flight and felt Mackay was overly dismissive of an issue that could have jeopardized the program. He and Mackay agreed that in the future if one or the other had reservations about something the other one was doing, they would hash it out instead of ignoring it.

Later, Moses told me that he was always confident about the feather saving the day but had lingering concerns about the source of the problems.

He asked me if I remembered Stephen Covey. I did not. In 1989, Covey wrote a book called *The 7 Habits of Highly Effective People*; it sold over twenty-five million copies.

"At NASA we had 'The 7 Elements of Flight Rationale,' building off the Covey craze. Basically, we had to answer seven questions to be ready to go and fly," said Moses.

One related to whether or not an identified problem was "bounded." If they knew the cause of a problem, it was "bounded." But if they didn't know the cause, and therefore how to avoid it, it was "unbounded." An unbounded problem? "You're not good to fly," said Moses.

That's what was worrying him most, that the engineering team couldn't immediately explain the aerodynamic forces playing on the spaceship that day: for the moment, the problem remained unbounded.

27.

DEFINING SPACE

IN LATE SEPTEMBER 2018 Stucky packed his Keurig machine, a case of Death Wish Coffee pods, and his tux. He and Agin drove to Disneyland.

They checked into an enormous Craftsman hotel with overpriced parking—the site of the annual Society of Experimental Test Pilots symposium. Stucky got a kick out of watching test pilots mingle in the lobby with sunburned dads in Goofy hats.

He and Agin went out for steak then returned to their room. She flipped through the TV channels, exhausted after a recent house-hunting trip to New Mexico. A move seemed inevitable. With three rocket-powered flights completed in four months, it felt like a matter of time before Virgin announced that it was uprooting and moving to Spaceport America.

Neither was thrilled about moving. Stucky liked being near L.A., a short drive from concerts and sporting events and, most important, from Dillon. (Sascha, his older daughter, had moved to Phoenix with her husband, Jonathan, and their two children, and another on the way, whom Sascha was homeschooling, while running her own business: an over-achiever. "I get that from my dad," said Sascha. She'd hired a "high performance coach" to help her "self-optimize" after realizing, "I was hitting my upper limits." Lauren, his younger daughter, was living in L.A. with her husband, with whom she had eloped.) Stucky and Dillon frequently went paragliding together, but honestly Stucky didn't much care what they did.

He just enjoyed being in Dillon's company, laughing at his jokes, hearing about his latest rehearsal—watching his son become a man.

Agin left the channel on *America's Got Talent*.

Stucky was lying next to her on the bed with his computer on his lap. He typed an email to Moses, appealing to Moses's good sense. Moses had seen the script Stucky was preparing to brief at SETP and told Stucky to water it down: more progress, fewer pitfalls.

Stucky objected. He knew his audience and what would and wouldn't pass with this crowd. Most of them were current or former military officers, serious types—technically fluent, operationally seasoned, attuned to bullshit and spin. Stucky wasn't about to step onstage and recite a list of program milestones that Stephen Attenborough or some PR person had prepared. That's not what SETP was all about. It was a community, a safe space, a forum where like-minded professionals could gather and share lessons they'd learned to try to prevent one another from dying doing a thing they all loved—flying airplanes into the unknown.

That's why I went, to see how test pilots talked among themselves, to hear what kind of questions they would ask Stucky.

Stucky made clear in his email to Moses that if he was ever criticizing the company, he was immensely proud to work there, proud of its mission, proud of his colleagues. Anyone, however, expecting him to be some corporate mouthpiece had gravely misjudged his character.

This was what often confused him about Virgin. Branson was swashbuckling and spontaneous, but it felt like Whitesides, the CEO, lived in perpetual fear of putting a wrong foot forward. This had been an ongoing source of dispute. In 2013, Whitesides tried to prevent Stucky from presenting at that year's SETP meeting. Stucky lobbied hard in favor. "If we do not present this year it will only encourage rumor-mongering that I feel is potentially more damaging to the program than giving a slightly sanitized version of the problems we've encountered and how we've learned from them," he said. "The potential benefits it has for aviation outweigh any perceived risks."

Whitesides relented.

But now, as Stucky went through his in-box with *America's Got Talent* in the background, he realized he had a more immediate problem.

In twelve hours, Stucky was supposed to lead a field trip to Mojave for a few dozen SETP members. He had been organizing the trip for months. He planned to show the group around FAITH and one of the Scaled hangars. Stucky was confident that his Mojave outing would upstage a previous year's SETP trip to Disney's Indiana Jones Adventure ride.

That was, until he saw the message from Scaled's HR department, informing Stucky that he was no longer welcome to lead the tour.

Annoyed, Stucky suspected the rescinded invite was retaliation for comments he'd recently made about Peter Siebold's leadership failings and cockpit performance on the day of the crash. Stucky made the comments in an article about him that appeared in the *New Yorker*, titled "Rocket Man." I wrote the article, based on almost four years of embedded access; hundreds of taped interviews; dozens of conversations with Stucky, Moses, and other key people; and a hard drive full of internal documents, company correspondence, cockpit videos, and more.

The article told a story about passionate people attempting to make an ambitious but troubled dream a reality. Curiously, the response to its publication revealed some of the reasons for that trouble—how a corporate culture obsessed with rosy news could undermine its own ambitions.

Two weeks after the story came out, Branson contacted me. "I'm rather wary of writing to thank journalists for the time and effort they've put in writing a piece and to compliment them, fearful that they may feel I'm stepping on their independence," he wrote. "But anyway, thank you for the enormous amount of time, effort and research you put in this article. It's rare these days." Additionally, he forwarded an email from Rutan: "Hands down, the finest article I have ever read about any New-Space company."

But the company line was less amiable. Stucky got invited to appear on a public radio program with me, but Whitesides declined sharply on his behalf. And while Moses was complimentary in private, he instructed his team to cease talking to me. "Please do not have any conversations with Nick about our program anymore," he wrote. "After being embedded for so long, it will be an active effort to pull back the access, but that's what we need to do."

This sudden, adversarial turn left me confused, until I learned that it was coming primarily from Stephen Attenborough. Attenborough had never liked the idea of allowing a reporter inside the company. He was a pitchman, after all; he saw reporters as marketing platforms. He was happy to get behind articles spotlighting Virgin's glitzy customers or newest sponsorship partner because he could craft the message in furtherance of Virgin's commercial interests. He thought my *New Yorker* article, however, made the program sound exceedingly dangerous and lobbied against extending the company's cooperation in support of this book without a financial stake in, or editorial control of, the book. These were terms to which I could not possibly agree. I was therefore on my own, trying to get back inside FAITH however I could—even as a stowaway on an SETP tour.

Stucky felt frustrated about getting shunned by Scaled, but he believed what he said about Siebold and didn't regret saying it. He was relieved that Scaled hadn't canceled the tour for everyone else.

Stucky would just have to wait outside—in the parking lot, on the bus he chartered, for the tour he arranged, at the company where he used to work.

THE NEXT MORNING Stucky was handing out bags of trail mix and water bottles to test pilots as they stepped onto the bus.

"Welcome, everybody," said Stucky, playing chaperone.

The driver turned on his mic and asked, "Is anybody prone to motion sickness?" None of the pilots raised their hand. The bus rolled out.

They had a two-hour drive ahead of them. Stucky put on a documentary about Burt Rutan and SpaceShipOne to get people in the mood. Rutan's witticism about the need of having confidence in nonsense drew laughs. Later, when one of Rutan's engineers bragged about Scaled never building the same vehicle twice, some gasped.

"They got balls," said the retired Navy pilot sitting next to me.

The driver made good time, so Stucky suggested we stop at the local doughnut shop in Mojave. "I'm buying," said Stucky, and we piled out.

"You a regular here, Forger?" asked Norm Howell, a retired Air Force officer with shaggy blond hair.

More like an investor, joked another pilot.

Howell ordered a chocolate doughnut and a Diet Coke. "Thanks for the doughnut," he told Stucky.

Howell went on. "There's an important parallel to space flight here, you know. The V-2 rockets had hypergolic fuel, right? You'd mix hydrazine and hydrogen peroxide together and wouldn't need an ignition source. This is the exact opposite," said Howell, holding his pastry and soda aloft. "It's a well-known fact that aspartame and sugar cancel each other out."

Stucky smiled. He asked whether everyone got a doughnut and, if so, to get back on the bus. "We need to keep moving," he said.

WHEN WE REACHED the Scaled hangar, Stucky waited outside. He didn't feel like getting into it with his fellow SETP members, so he said he had to catch up on some email.

Afterward, Stucky showed the group around the new rocket test site and then led us to FAITH. "Our last stop," he said. Stucky provided an overview of the program to date and said they were currently stripping and "fine-tuning" the thermal protection system on SpaceShipTwo because they were planning to fly higher than ever on the next flight and didn't want to overheat on reentry.

Stucky divided the group into smaller ones. Mackay led one group to the simulator, which is like a large gaming pod with a spaceship cockpit, and then narrated what Virgin hoped would be a typical space flight. As the ship reached its apogee, Mackay said, "Those little puffs you hear? That's the RCS."

Howell raised his hand and said he noticed that the altimeter at apogee had peaked at 318,000 feet. "Aren't you trying to get to three twenty-eight?" he asked.

According to a leading international aerospace body, the Fédération Aéronautique Internationale, or FAI, space begins at 328,000 feet above sea level. The X Prize had used this boundary, as did Virgin's main competitor, Blue Origin.

Blue Origin had continued to make progress with its suborbital rocket. *Step by Step, Ferociously.* Two months earlier, in July 2018, their New Shepard went 389,000 feet above sea level with a test dummy—"Mannequin Skywalker"—on board. Bezos tweeted, "The lucky boots worked again."

Mackay told Howell that Virgin was "trying to get as high as we can." But he seemed to be choosing his words carefully, perhaps because he knew what company insiders, but few outsiders, did: that Virgin wasn't aiming for three twenty-eight anymore.

Or at least not yet.

Because of all the mods and reinforcements, SpaceShipTwo was much heavier than Virgin had initially intended, and they didn't think the ship could reach three twenty-eight. They had already reduced the number of passengers from six to four.

Now they were even adjusting their definition of space.

William Pomerantz, an affable Harvard grad and longtime VP of "special projects" at Virgin, had been out discussing the boundary of space with scientists, including an astrophysicist at his alma mater named Jonathan McDowell. Space was fungible and there was nothing magical about three twenty-eight, Pomerantz contended.

Hearing about these discussions made Stucky awfully uncomfortable at first: it felt like Virgin was trying to change the rules in the middle of the game. SpaceShipOne had flown to three twenty-eight, and Scaled designed SpaceShipTwo with the same goal in mind, so it felt cheap to be not only lowering their sights but, furthermore, trying to persuade scientists that their math was wrong. Gaslighting astrophysicists seemed a foolish gambit.

But Pomerantz was onto something.

How *could* you define or delineate space? No one had even bothered before humans began trying to go there. In 1957, an American named Andrew Haley, the president of the International Astronautical Federation—unrelated to the FAI—proposed a "critical jurisdictional line" 275,000 feet above sea level, where "airspace" ended and "outer space" began—an imaginary line "separat[ing] the territory of air-breathing vehicles from that of rocket vehicles." Citing the research of a

Hungarian-born physicist, Theodore von Kármán, Haley called his line the Kármán line.

A year later, the FAI convened a group of American and Soviet scientists, who came up with their own space boundary: one hundred kilometers, or about 328,000 feet.

Haley, curiously, acknowledged the FAI's boundary; he said it "coincided" with his own, despite the 53,000-foot discrepancy. The Kármán line and the FAI had been more or less synonymous ever since.

In October 2018, McDowell published an article in *Acta Astronautica*, a peer-reviewed monthly, titled, "The Edge of Space: Revisiting the Karman Line."

McDowell drew on history, explaining how, in the late fifties, the Air Force began awarding astronaut wings to pilots who flew over fifty statute miles, and how fifty miles was not only "a nice round figure" but also "the right choice from a physical point of view" because the mesosphere started about fifty miles above sea level. (The Kármán line was sixty-two miles above sea level.)

McDowell mounted a scientific argument, too. Like von Kármán, McDowell contended that our notion of space should begin wherever orbital dynamics exceeded aerodynamic forces, but McDowell demonstrated how, based on ballistic coefficients and modern atmospheric models, fifty miles was a "suitable choice to use as the canonical lower 'edge of space' in circumstances where such a dividing line between atmosphere and space is desired."

Mackay didn't bore Howell with any of these details. He stuck to the script: "It would be nice to get to three thirty."

THE NEXT MORNING Stucky strode into a hotel ballroom wearing a tie, a Virgin-branded button-up, and a black leather bomber. He grabbed an open seat and took notes during the other presentations: Bell pilots discussed a tilt-rotor aircraft they were testing for the Army, and Air Force pilots described the novel refueling techniques they were developing for the C-17.

Everyone had their niche.

Stucky and Agin went out for lunch with an experimental hang-gliding pilot. Stucky and Mackay took the stage after lunch.

Stucky reminded the audience that he and Mackay had previously shared some of the lessons they had learned from the accident, and now they wanted to talk about what they were learning from their new space-ship. Stucky admitted to encountering some "potholes" but said he was confident that they had addressed the problems and now had a better ship.

Stucky played some of the footage from the first powered flight. "All was fine until we got past Mach 1.5, where I started having roll chal-lenges," he said. "I put in a hard left roll boost command and the roll rate was slowing down nicely, but suddenly the gyro flipped."

"We were inverted at eighty-five thousand feet," said Mackay. "Some-thing even Forger had never done before."

This got some laughs from an audience of extraordinarily accom-plished men and women, whose rapt attention suggested that, even among them, Stucky was something unique.

But not everyone was impressed.

Todd Ericson stood in the back of the room, seething. He complained about Stucky "rewriting history," grumbling about how the gyro malfunc-tion was "real, but secondary." Stucky did not have to abort because of a software issue, said Ericson, but "because he was losing control."

28.

DAGOBAH

WHILE STUCKY WAS IN DISNEYLAND, a technician back in Mojave spotted a bond line oozing from a seam on SpaceShipTwo's left tail boom. The spaceship was held together by fancy glue, so the sight of some dried, excess glue was not an immediate cause for alarm. That's why they had pencil-grinders—to grind down the ooze. But when the tech powered up his pencil-grinder and pressed on the ooze, which was normally dense, it flaked like meringue.

Now there was cause for alarm.

Flaking was a symptom of a porous bond and a porous bond was a weak bond, and a weak bond was prone to come unglued, which—when they looked closely—they realized was already happening.

Moses despaired when he heard the news. He had a hunch that the problem wasn't isolated, that this would be more than a few days' setback. In a rare show of frustration, he told the engineer who broke the news, "I hate you."

The team diagnosed the problem and traced it back to a bad batch of glue. It was as simple as that. Whoever had mixed that particular batch on that particular day years earlier had waited too long before slathering the bond onto the skin, so that when it cured it did so with a filmy layer on top, a phenomenon known as "blushing"; Stucky liked to describe it as the film that forms on mushroom soup left on the stove.

It made Stucky uneasy. He was okay flying an imperfect ship if he knew where the imperfections were. But on his spectrum of real versus imagined risks, a bad bond qualified as very real. He shuddered to think about charging through the atmosphere at twice the speed of sound if suddenly one of those air bubbles in the bond popped and caused the boom to shear off.

Disbonding wasn't an unforeseen problem. Virgin had encountered bonding difficulties in the past. "Heads up," Moses had informed the management team, back in 2013. "We are finding some deeper issues, including the upper wing skin being debonded in places from the wing spar." Those repairs grounded them for months.

This, like that, was not a job worth rushing. Stucky suggested the engineers test every bond on the ship by slapping industrial-size suction cups on the outer skins and then trying to pull them off. His recommendation was deemed indelicate. Instead, the maintainers cut panels into the flawed boom, so, with the aid of borescopes, the techs, working blind in narrow crevasses with sometimes only an arm inside the ship, could track their progress on a flat-screen monitor. *What irony*, thought Stucky, *after all their high-tech aerodynamic and propulsion challenges, to be grounded because of a batch of subpar glue.*

This was going to take months, he knew. In early November, he and Agin drove to northern Idaho to spend a couple of nights with Burt Rutan at his home on the banks of Lake Coeur d'Alene. Though Rutan was getting old, he remained arguably the most innovative aerospace designer alive. Stucky's trip felt like Luke Skywalker going to see Yoda in Dagobah.

Stucky asked Rutan whether I could come along. Rutan agreed.

STUCKY AND AGIN and I all arrived on the afternoon of November sixth, the day of the 2018 midterms. A yellow DON'T TREAD ON ME flag flew from a pole in the front yard.

Rutan welcomed me inside. "You're in one of the most conservative counties in the United States," he said. Ruby Ridge was just up the road. I asked about the fires I saw on a piney ridgeline driving in. "That's where we burn the liberals," said Rutan.

From left: Mark Stucky, Burt Rutan, and the author in Rutan's hangar, in 2018.

He and his fourth wife, Tonya, were hosting an election party that night. There were Crock-Pots on the kitchen counter warming rice and beans and pulled pork. The house had a chic lodge aesthetic, with high ceilings, a central stone fireplace, broad views of the lake, and a globe on an end table, though Rutan did not believe the globe was warming. "Biggest science fraud in history," he told Newt Gingrich in an email. Gingrich had contacted Rutan after Donald Trump became president, to solicit Rutan's ideas for how Trump could rejuvenate NASA, which Rutan provided, along with some unsolicited ones.

As for NASA, Rutan maintained his dim assessment of the agency. He rated its performance poor, not because of any particular program but because of its failure to inspire. That should be its lodestar, each line in its budget scrutinized on the basis of whether or not that item served a bigger, inspirational goal. "It's hard to define something more important to America's future than to assure that schools are inspiring kids," Rutan told Gingrich. The problem with NASA? Too many bureaucrats, not enough breakthroughs. "To have any chance of escaping the deep rut we have been struggling in for nearly 4 decades we will have to dramatically change how research is funded, managed, and regulated," he wrote. "Success depends on draining the swamp."

I followed Rutan downstairs into his trophy room. He was showing me the Presidential Citizen's Medal he received from Ronald Reagan ("Obama handed them out like popcorn, but before that it was really a prestigious thing"), his two Robert J. Collier trophies, and the solid gold medallion he got from Aero Club de France ("Identical to the one they handed Lindbergh in 1927"), when his guests began to arrive.

Rutan introduced me, saying that I'd just flown in from Washington, DC.

"You have any gator bites from living in the swamp?" one guest asked. An hors d'oeuvres platter went by.

The party had no sooner begun when returns started trickling in from the East Coast. It looked like it would be a bad night for Republicans. Rutan turned away from the TV and said, "Whether I watch it or not it's going to be the same," so he and Stucky and I prepared plates of fajitas and went upstairs, into Rutan's office.

Rutan popped a quiz: he asked Stucky if he knew how many different rocket systems had put humans into space during the 1960s. "I've given this test to several NASA administrators," said Rutan. "Nobody's passed it."

Stucky took a shot. There was Vostok, which flew Yuri Gagarin in 1961.

And Redstone, which flew Alan Shepard the same year.

Then there was Atlas, which flew John Glenn in 1962.

And obviously the X-15.

Titan II powered the first Gemini mission in 1965.

"You're forgetting some of the other Soviet rockets," said Rutan.

Voskhod flew in 1964, and Soyuz made its first manned space voyage in 1966.

And don't forget about Saturn IB, which flew Apollo 7 in 1968, and Saturn V, which flew Apollo 11 in 1969. "The first manned space flight of Saturn V went to the fucking moon!" said Rutan. His eyes lit up with awe. "One more!"

Stucky threw up his hands, and when Rutan looked at me I shook my head.

Rutan said, exultantly, "The upper stage of the LEM!"—the lunar module that Neil Armstrong and Buzz Aldrin landed on, and launched from, the moon. "It was sitting on the moon and it put astronauts into space."

Unconvinced, Stucky said, "They were already *in* space."

"No," said Rutan. "They were on the moon. If you're on Mars, are you in space?"

"Mars has an atmosphere," said Stucky. "The moon doesn't."

Rutan paused. "Good argument. Tonya thinks it's a trick question, too." But he shrugged it off: he had a point to make about innovation and inspiration. He asked if we knew how many new rocket systems had flown humans to space in the fifty years since then.

"Three!" he said. "Shuttle. Chinese Shenzhou, in 2003. And Space-ShipOne. I'm not proud of that. I should be. But how can you be proud of that fact that almost nothing happened for forty-eight years, after ten new systems in nine years? Isn't that a tragedy?"

WE WENT OUT the next morning to a restaurant renowned for its huckleberry pancakes. Over breakfast, Stucky told Rutan about the bonding crisis they were dealing with back in Mojave.

"What are you bonding to what?" asked Rutan.

"Carbon to carbon," said Stucky.

"And you're using a 3M paste? We didn't even use that on Space-ShipOne."

"That explains everything," said Stucky, shaking his head. He was tired of hearing Virgin's engineers always dumping on Scaled and insisting they knew some better way when Scaled had been using a simpler, and cheaper, method already on the shelf.

Rutan said, "We used what the homebuilders use: flox"—or flocked cotton. "You know what flox is? It's belly button lint. You want to mix it so there's a little shine left in it. There weren't any blushing issues. We didn't have any of those problems until we tried the fancy shit."

He told Stucky how he had recently gone down to Mojave. He'd seen the two new spaceships Virgin was building—SpaceShipTwo models three and four—and was surprised Virgin was doing that when they hadn't finished expanding the envelope of the current one.

"You must have me confused with the program manager," said Stucky. He described some of the modifications they were making on the next two ships, such as adding a stability augmentation system to the h-stabs to help keep the wings level around Mach 2.

"We didn't have stability augmentation. Or a yaw damper. The control system on SpaceShipOne was the same as a Piper Cub—cables for rudders and push rods to the elevons—but we did have electric trim for the stabilizers. It had to be like that: we didn't have enough time. First time anyone's gone supersonic in an airplane without wind tunnel data," he said. His team was good. But, "You could argue that we were damn lucky."

Stucky groaned about how the ship was getting heavy, and how some engineers were insisting that they apply more thermal protection to the leading edges of the wings and booms. Rutan recalled having these same arguments with his thermal engineer, who had claimed Scaled needed forty-two pounds of thermal protection on SpaceShipOne. "I talked him down to fourteen," said Rutan. "We needed four."

How much did they have on SpaceShipTwo?

"A couple hundred," said Stucky.

Rutan had heard this story too many times before—risk aversion run wild, perfection becoming the enemy of good. Engineers were naturally fixers, and he had seen talented, yet risk-averse, ones muck things up with endless tweaks and improvements and modifications, he said, "Right up until the point when a program runs out of money."

RUTAN BUILT SPACESHIPONE for about $25 million, and after the craft completed three space flights and won the $10 million X Prize, Paul Allen, by selling the patents, even eked out a profit.

But SpaceShipTwo was becoming a money pit. Fourteen years and four casualties and no space flights. In 2018, Branson estimated having spent nearly $1 billion on the program. Customers defrayed some of those costs with their $200,000 reservations, and an Emirati fund put up about $380 million, which certainly helped. But the program was bleeding cash. In 2018, they spent $118 million on research and development expenses, earning just $3 million in revenue.

They explored other ways to raise money. Ericson was pushing for them to go after government contracts. Government money was good money, he told Whitesides—slow, but lucrative and reliable. Ericson, who was old friends with the Air Force chief, David Goldfein, flew to DC to brief Goldfein and his staff on what Virgin wanted to do, including hypersonic, point-to-point travel. The generals sat up and listened. Later, Goldfein talked about how he could envision the day when US Navy SEALs and Delta Force operators could hop on some version of SpaceShipTwo and be "any place on the planet in less than an hour." Still, the Pentagon wasn't ready to cut Virgin a check just yet.

Branson had his eye on another potential funder—the crown prince of Saudi Arabia, Mohammad bin Salman. Branson had recently flown to Riyadh, where bin Salman pledged $1 billion, and when bin Salman came to the United States, Branson hosted him in Mojave, showing the prince around the hangar. A billion dollars would ease a lot of Branson's worries.

But then Jamal Khashoggi walked into the Saudi consulate in Istanbul and never came out. Investigators concluded that Khashoggi—a reporter and US permanent resident who wrote critically of bin Salman in his *Washington Post* columns—had been killed and dismembered with a bone saw on bin Salman's orders. Branson, a human rights advocate, refused to accept money from someone complicit in such a gruesome act. He walked away, leaving $1 billion on the table.

Stucky wasn't one for corporate cheerleading but felt proud of Branson's decision. "I'm heartened to see Sir Richard demonstrate so publicly that, despite immense riches, a state should not be able to buy all their desired friends if they choose to not obey the most basic of human rights," Stucky wrote to Whitesides.

Now they needed to find $1 billion from somewhere else because, as Stucky liked to say, "No bucks, no Buck Rogers."

AFTER BREAKFAST, RUTAN drove across town to show Stucky the progress he had made on the SkiGull. Or rather the lack thereof. Looking back, Rutan said he should have given up on the SkiGull years ago, but now he was hostage to sunken costs—not unlike Branson's own predicament.

"I can't quit now," said Rutan.

Stucky suggested we go watch *First Man*, the new movie about Neil Armstrong.

"I've heard it's junk," said Rutan. Fox News hosts had been frothing about how the filmmakers didn't show Armstrong and Buzz Aldrin planting the US flag on the moon. "It's almost like they're embarrassed at the achievement coming from America," Donald Trump said of the film. Still, Rutan agreed.

We got popcorn and found our seats. During the opening scene, where Armstrong flies an X-15 to the edge of the atmosphere and loses aerodynamic control and has to make an emergency landing on a dry lake bed at Edwards Air Force Base, Rutan leaned over and whispered to Stucky, "The joys of suborbital flight."

Rutan kept up a running commentary for the next two hours, a shushable offense. If only the other theatergoers knew that the babbling septuagenarian in the front row was the brains behind the last novel rocket ship to transport humans to space, and that the man sitting next to him might well become the next astronaut.

"Holy bananas, this is fantastic!" Rutan declared as Armstrong and Aldrin's LEM touched down on the moon. When a few minutes later the LEM launched off the moon on its way to rendezvous with the Apollo capsule, Rutan, citing his trivia quiz, said, *Number ten.*

WE GOT BACK to Rutan's house and found a kid from the neighborhood at the door. Tonya and his mom had made arrangements for the student to come over and speak with Rutan about his science project. "He wants a rocket scientist to comment on whether it's any good," said Tonya.

Tonya introduced him as Grant, an eighth-grader from the local STEM school. Rutan had spoken at the school when it opened, and the school continued to attract impressive guest speakers from the area. A former astronaut had recently come to talk about his three space missions; too bad he didn't bring his space gloves, said Grant.

"He's frantic about somebody stealing them," said Rutan, who knew the astronaut.

Grant asked Tonya for a glass of water, then opened his notebook and took out a pen. He was entering a robotics competition and hoping to find a way to alleviate the excruciating headaches that astronauts called "pumpkin head." Grant proposed an artificial magnetic field around the International Space Station, imagining a plexiglass-enclosed ring of water that would encircle the ISS, and, through solar energy, create a magnetic field.

Rutan thought for a moment. He admired the ambition of a young, uninhibited mind. But about this artificial magnetic field? Rutan asked Grant how he thought magnets would impact human blood flows, when magnets had no effect on blood, and then digressed into a meditation on magnetism. "Did you know that Velcro is the only fastener that improves with vibration?" he asked.

Stucky interrupted Rutan, on Grant's behalf. "This doesn't solve your initial thing, which was pumpkin head," said Stucky.

Rutan apologized for failing to make a proper introduction. He told Grant, pointing to Stucky, "The guy sitting in that seat is one of the most famous and capable test pilots ever."

Grant lit up. "One of the big things they look for when judging your project is how much research you did and who you interviewed," he said. "And it will be really good to say that I asked multiple smart people about the topic, backing it up."

Maybe, said Rutan. But just because they were talking about this didn't mean Rutan thought it was a good idea.

"I don't say things are stupid ideas, because before you have a breakthrough, it's normally called *nonsense*. Confidence in nonsense is what's needed. It's the stupid people that have the stupidity or the courage to try nonsense, and therefore they get the breakthroughs," said Rutan.

"About your pumpkin head thing? Working magnetic fields to pull blood out of your head? Sounds like nonsense to me."

29.

FORTUNE COOKIES

STUCKY CANCELED THANKSGIVING. He hated doing it, but he had been going nonstop since he got back from Idaho, and so he called his kids a few days before the holiday and begged for forgiveness. He promised they would gather soon, after his next flight—the big one: he was going to space.

Stucky bought a chateaubriand from Costco that he and Agin ate on the couch, watching Thanksgiving Day football. Plates, wineglasses, and books covered the coffee table. Stucky flipped through books about rocket ships during the commercials.

He had always brought work home. He felt like he owed it to the Marines, to NASA, and to the Air Force, in exchange for the "trust and confidence the nation puts in you to be the tip of the spear." Plus, it seemed obvious that if you wanted to be great at something, you had to work for it. People made excellence more mysterious than it had to be. Sometimes, said Teller, the magician of Penn & Teller, "magic is just someone spending more time on something than anyone else might reasonably expect."

Stucky had read all of the books that he planned to read that weekend at least once, old X-15 books like *X-15: The NASA Mission Reports* and *Hypersonic: The Story of the North American X-15* and *X-15: Extending the Frontiers of Flight*. But when he read them the first time he was young and in for the titillating accounts of hypersonic flight. Now he was looking for clues. Stucky knew the X-15 had its share of problems with sideslip and lateral instability; he was rereading the books to see what he could learn about how they fixed it.

Over the weekend, Stucky did his research, and on Monday he emailed Bob Hoey, one of the X-15 engineers. Stucky asked Hoey for "some specifics on the beta dot technique"—a tactic the X-15 pilots used to keep the airplane steady in turbulent moments. "I understand it was a quick lateral stick input when the needle was passing through zero but that's not enough info for me."

"You are taxing my memory," said Hoey, who was now in his eighties. But Hoey was being modest, as he proceeded to give a deep technical recitation of the relevant details, about how the X-15 wanted to slide and roll whenever it was descending at an angle of attack of fifteen degrees with the roll damper off. He told Stucky that the test pilots on the program had been experimenting with different rudder techniques that were "completely ineffective," before engineers discovered the "beta dot technique," which involved a "sharp aileron pulse, applied as the beta needle was passing through zero." Somehow, it stopped the oscillation.

Later that day, Stucky tried this technique in the simulator. Sadly, it didn't work. But Stucky found another technique that did. It was all about how he gripped the stick: he typically squeezed the grip and made whatever twitchy minor adjustments were needed. But he found that the controls were so sensitive approaching Mach 2 that his inputs were doing more harm than good; if he opened his hands flat and held them on either side of the stick, like bowling alley bumpers, he could let the stick float and it actually limited what pilots called PIO, or the pilot-induced oscillation. The technique was known as "dead-banding." He told Hoey.

"Looks like you are on the right track," Hoey replied.

STUCKY CARRIED ON. The flight was just over a week out. The ship was ready; they had stripped the bad bond from the left tail boom and reapplied a new one. The long-term forecast looked good. Stucky had a lengthy punch list, but it was manageable.

He dug through his closet for an old T-shirt. He acquired the shirt years ago, while he was doing secret test work in Nevada. After Stucky flew the first flight of an experimental jet, colleagues sprayed him with champagne and gave him the shirt. It was an old, green T-shirt with a four-leaf clover on the front, and LUCKY written across the back. Stucky donated the shirt to a bar on the base out there in the wasteland where all the test pilots hung out. The bartender tacked the shirt on the wall. When Stucky was getting ready to retire and move to California to work for Scaled, the bartender took the shirt down and told Stucky, "Fly this to space and give it back."

Stucky found the shirt and pouched it in a manila envelope that he would tuck under his seat in SpaceShipTwo. Others were vying to get their own tokens and tchotchkes on board. Stephen Attenborough saw an opportunity to market the brand, and since it was almost Christmas he proposed that Stucky bring a sprig of mistletoe with him. The mistletoe would float around and depict microgravity; Stucky joked he could make a funny moment out of it by pretending to kiss his copilot.

This went over poorly with Stucky's copilot, C.J. Sturckow. Sturckow was cranky. Once, on his lunch break, Sturckow went jogging in the middle of a Mojave sandstorm and when he came back, Dave Mackay, the Scot, asked how it felt running in that wind. Sturckow went off about how the marines who fought and died in Vietnam never complained about the wind and how he wasn't about to start.

As for the mistletoe? Stucky wasn't about to let a thorny, untethered item into the cockpit, and Sturckow wasn't about to kiss Stucky. Mistletoe was out.

What about a bell? They were festive, and a nod to the first album Branson produced in 1973, Mike Oldfield's *Tubular Bells*. But no one wanted a metal object loose in the cabin. Bells were out.

How about a miniature E.T. doll? Too ridiculous, said Moses. E.T. was out.

A stuffed Richard Branson doll? They rush-ordered a few but they looked like voodoo dolls. "Creepy as shit," said Moses. Dolls were out.

They settled on a paper snowflake. Attenborough and his marketing team had another ask. As they thought about all of the magical things

Stucky could say from space, they proposed giving Stucky some talking points. Or perhaps Stucky could prerecord something that Virgin's multimedia wizards could later splice into the video footage.

Absolutely not, said Stucky.

He told Mackay to tell Whitesides and Attenborough that he swore not to say anything embarrassing and that he was determined that the audio from the cockpit would be fit to broadcast in an unedited form. But no way was he going to parrot somebody else's lines.

"I'm okay risking my body for VG," he told Mackay. "But I'll be damned if I'm going to give them my soul."

AT A BRIEFING on Monday morning, seventy-two hours before the flight, Moses felt compelled to counsel calm. Nerves were normal, he said. He felt them, too. But, "If you're going to have your fifteen minutes of freakout, let's get it over today."

Stucky and Sturckow rehearsed in the simulator a few more times. They didn't share the same kind of frank and open relationship Stucky had with Mackay, but then again you didn't have to be best friends to complete a mission. Michael Collins would describe his relationship with Neil Armstrong and Buzz Aldrin on Apollo 11 as one of "amiable strangers," but they trusted one another in the cockpit and that was what mattered.

Besides the new dead-banding technique, Stucky planned to try something else on the next flight that he had never done before. He was going to drop from WhiteKnightTwo at a lower altitude than usual. This struck some as nonsensical: If they were trying to fly *higher* than ever before, why would they want to start *lower* than ever before?

Hear me out, said Stucky. A couple of crack engineers on his team had done the math. Stucky thought the world of these two: one of them, Brandon Parrish, was about to propose to his girlfriend; Stucky had agreed to take Parrish's engagement ring into space. According to Parrish and his colleague, SpaceShipTwo's lateral instability problems around Mach 1.5 were not necessarily because of its airspeed but because of its airspeed *at a certain pitch and altitude.* They believed that if WhiteKnightTwo dropped SpaceShipTwo lower, into thicker air, the spaceship could achieve a more stable pitch attitude—or relative orientation—on its way to Mach 2.

Everyone agreed that it was a worth a shot.

On Tuesday night, thirty-six hours before the launch, Stucky and Agin went out for dinner at the Broken Bit. Astronaut nutrition was not as restrictive as it had been in the early days, when John Glenn was eating cubes concocted in a laboratory for maximum nutritional value and minimal fecal production. Stucky was nonetheless careful about what he put in his body. He ordered an ahi tuna salad but asked the waitress to hold the oil because oil sometimes messed with his stomach. And though Agin had a glass of prosecco, Stucky went with water. "I'll probably screw something up because I'm not drunk," he said.

His dietary regimen was not the only thing that distinguished this flight from the early astronaut missions. Those were big, televised events. Statements of national pride and intent. Occasions that dangled the prospect of some new existential reality. Hundreds of millions of people would watch from around the world, wondering what those brave astronauts were going to find on the other side—and if they would make it back to tell the rest of the world what they saw.

A part of Stucky yearned for that kind of fanfare—photographers on his lawn, his words enshrined on Christmas ornaments, autograph sittings, exclusive magazine deals, random people bidding to own locks of his hair.

But he knew the world had changed. People were more cynical now. Politics had spoiled everything, even heroism. He wondered how John Glenn and Neil Armstrong would have fared under today's media scrutiny, in a world where it seemed like only fools believed in heroes anymore.

None of this made his mission any less risky, though. Who knew what he was going to find on the other side of the envelope. And what if he didn't come back?

We had discussed this. I asked Stucky what would happen if he had a bad day. Would he consider giving me permission to go through his stuff? He thought about it, then agreed. He told me that Agin knew all of his nonwork passwords, and that if something tragic occurred she would help me get into his accounts.

We are often asked as authors how we find our stories, how we know that a subject will hold our interest over the years required to research, report, and write about it. Truthfully? We don't. We are in that sense little different from test pilots or gamblers, taking risks on an uncertain future. We gather whatever information we can and hope that it leads us to an interesting end.

I had no idea when I began this project that Stucky knew my father. Or how much they would remind me of each other, how writing about Stucky would become an opportunity for me to write about my dad without writing about my dad, because I knew that my father would never let me get close enough to write about him in a way that felt revealing, even perhaps discomfiting. That was an awkward reality to confront, how it was that I could ask Stucky questions that I had never asked, and might never ask, my dad.

That was when I knew that I could spend years writing this book.

But we, the authors, are not the only ones doing the choosing.

Sometimes subjects choose their authors. At one point Stucky asked me why I thought he had done what he had done—cooperating with me, letting me into his home and into his life, exposing himself to a reporter.

I said that vanity probably played a role, and he wasn't too vain to admit that. But there was more to it, and I suggested that he knew he had lived a remarkable life and had been waiting to find someone in whom he trusted to tell it, and that when I walked into his life he knew I was that person. He agreed.

Stucky went to work the next day as usual. He dipped his hand into a bowl of fortune cookies and pulled out one that read YOU WILL CON-TINUE TO TAKE CHANCES AND BE GLAD YOU DID. He wondered whether someone had stocked the bowl with good fortunes, like Saddam Hussein had stocked those ponds around his palace with enormous bass. He left FAITH around three p.m. and drove to the house he owned and had been renting out to his stepdaughter. She had recently moved, and Stucky was preparing to put the house up for sale. He spent an hour hanging window blinds.

When he finished with the blinds, he met Agin at home. They ate an early dinner and had a relaxing soak in their hot tub. Before Stucky got under the sheets, he sat on the edge of the bed, closed his eyes, and chair-flew the mission one last time.

30.

THOROUGHBRED

I WOKE UP THE NEXT morning at four and saw two messages on my phone from Stucky. He sent the first at 3:29: Agin "came totally unglued. Doesn't want me to fly. What should I do?"

I sat up and rubbed my eyes.

"Yes, I am totally joking," read the second.

He had been up for an hour and was feeling rested and alert. A bout of butterflies had come and gone. He got dressed in his office to avoid waking Agin. Having stepped into the legs of his flight suit and pulled the zipper up to his neck, he double-checked the pocket along his left biceps for the jewelry box with Parrish's engagement ring, another one for his wedding band (which didn't fit him, and which Agin wore on her thumb), and the pocket along his calf for four Earth-shaped stress balls. He ate a cup of yogurt on his way out the door.

Stucky drove to Mojave in silence. No "Rocket Man" this time. It was a clear, predawn morning.

Venus shined bright, and a comet arced across the southwest sky. Stucky reached FAITH at four a.m. and went through his now-familiar drill—he and Sturckow flew the simulator with the latest weather balloon data, took a shuttle to the end of the runway, climbed into SpaceShipTwo through the hatch, and buckled in for the ride. Stucky reached under his seat to feel for the LUCKY shirt in the manila envelope: it was right where he put it.

The sun rose a Creamsicle orange on the eastern horizon, as hundreds of parka-clad spectators—employees, FAA officials, ticket-holding customers—gathered alongside the runway. Some huddled in heated tents. Agin and the pilots' spouses arrived in chauffeured white Land Rovers. Roadies had built a stage and set up a massive screen for the event.

Whitesides was pacing anxiously. He knew that failure was most definitely an option, as behind him loomed the abandoned hangar of Rotary Rocket, one of several once-promising, but now-defunct rocket companies in Mojave. Dark, underslept circles rimmed his eyes.

Branson walked onstage to welcome the crowd. "I'm not allowed to say it, but hopefully we're going to space today!" He turned to Whitesides. "Sorry, George. I know, 'It's a test flight.'" Maybe they would get all the way to space. Maybe not. Still, said Branson, "Hopefully we'll have some magic in the next couple of hours." He was skipping a wedding in Mumbai for the daughter of India's richest man to be here.

Branson felt he was on the cusp of achieving something very big. He could barely contain himself. The previous afternoon he had toured the test site run by his other rocket company, Virgin Orbit, the satellite launch provider which Branson spun off from Virgin Galactic in 2017. He wandered around, mouth agape at the Asimovian scene of engineers bent over intricate foil-wrapped tubes and hoses and piping, multistory tanks filled with oxidizer, industrial scaffolding rigs, floodlight-capped lampposts. And *he* was the one behind it all. *His* vision. *His* dream.

Branson had asked one of the Orbit engineers how many more ground tests they intended to run before they could launch an aerial test. At least a couple, said the engineer.

"Stop testing!" said Branson, half-kidding. "You might find something wrong!"

AT ELEVEN MINUTES past seven, WhiteKnightTwo accelerated down the runway and lifted off.

Stucky, sitting below in SpaceShipTwo, consulted his checklist as they made their ascent. Cabin pressure looked good. Their flight path was on course. He and Sturckow inhaled pure oxygen to improve their chances of survival, in case they had to bail; the dense oxygen amplified Stucky's breathing, reminding him of Darth Vader. As WhiteKnightTwo hooked around over Death Valley to head back toward Mojave, Stucky rehearsed the boost portion once more in his mind.

They reached the drop point, 43,000 feet above the desert. The White-KnightTwo pilot pushed the release button. *Thump.* SpaceShipTwo dropped away.

"Fire!" said Stucky.

Sturckow pulled the rocket motor switch. Clean light. They were off.

Stucky held the nose flat, according to plan, but once they broke the sound barrier, he began trimming the horizontal stabilizers and pitching into the gamma turn. Now the spaceship was vertical.

Sturckow relayed their trim readings, indicated by their degree of angle of attack:

"Seven . . . Eight . . . Nine . . . Ten . . . Eleven . . . Twelve . . . Thirteen."

Ten seconds.

"Good trim," said Stucky.

Twenty seconds.

"Copy," said Sturckow. "One-point-four Mach."

Stucky felt good about his open-hand grip and their pitch attitude. The light was dimming outside. Everything was looking fine.

Thirty seconds.

Stucky checked the altimeter: 70,000 feet. Based on their calculations, if they burned the motor to 135,000 feet, they would coast beyond their target of 264,000 feet. He just had to hold on.

But Stucky felt the wings rolling right; a single switch on the stick controlled pitch and roll, and when Stucky was trying to adjust trim he accidentally thumbed roll. A yellow caution light flashed on the console: the ENTRY GLIDE CONE warning. They were veering off course, but Sturckow knew Stucky could fly through it.

"ENTRY GLIDE CONE—ignore," Sturckow said. "Two-point-oh Mach."

Forty seconds.

Stucky rolled left to try to level the wings, but he overcompensated and rolled all the way over—86,000 feet up, going twice the speed of sound. Stucky held his nerve and let his input play out; the spaceship was still going straight up. He knew from all of his time in the simulator that there was no use fighting the roll.

A week earlier Stucky had pulled off a feat in the simulator that dumbfounded even Moses. It was during an emergency protocol session; the sim conductor gave Stucky a nasty RCS scenario: feather up; primary system failure, with one thruster stuck on; secondary system failure, with one thruster not working. The ship began to swirl and swivel, as if it was out of control. Moses assumed that Stucky was done, that there was no way Stucky could recover from a high-altitude departure with two failed RCS systems. But Stucky had other ideas, and though his RCS wasn't working properly, it was still working, and if he could get one burst of thruster from the primary system, followed by one burst of thruster from the secondary system, he figured that he could eventually regain control. "It was brilliant," said Moses. "He alternated between the failed one on and the failed one off. He turned it on for a second, *bam!*, fired it, and killed the rate, and then turned it right off. When the nose moved he turned the other one on and, *bam!*, kicked the nose over, and then turned it back off. He took what was a complete nose-and-roll departure and brought it around for a perfectly stable reentry. We're in the control room and I'm going, What the hell was *that*? He was moving in three axes and he had basically two options. He made the right decision every time"—within a matter of seconds. "It's like a sixth sense for him."

And now was no different. Rather than fighting the roll, Stucky let the ship take the lead.

"Bringing it back around," he said.

"Roger," Sturckow replied.

They were inverted, shooting for the stars.

Fifty seconds.

They sliced through the thinning air. "Getting good," said Stucky. The rocket motor was burning about two hundred pounds of nitrous oxide and rubber a second, becoming lighter and faster. "She felt like a thoroughbred," Stucky said. They were in unknown, uncharted aerodynamic territory, but Stucky never felt more sure of anything in his life.

"Two-point-eight Mach," said Sturckow.

"Copy," said Stucky, a lilt suddenly detectable in his voice.

Sixty seconds.

As Sturckow shut down the motor, Stucky checked the altimeter—135,000 feet—and glanced at a reading on the console that showed their estimated final altitude: 275,000 feet. Good attitude. Minimal roll rates.

Stucky got on the radio. "Great motor burn, everybody," he said. "We're going to space, Richard!"

BRANSON WAS SQUINTING against the sun, tracking SpaceShipTwo's contrails across the blue morning sky. From a dais onstage, Enrico Palermo, the president of The Spaceship Company, Virgin's manufacturing subsidiary, was calling out the spaceship's altitude.

"Two hundred thousand."

"Two hundred and twenty thousand feet."

"Two hundred and forty thousand feet."

Tears welled in Branson's eyes—a dreamer living his dream.

"Two hundred and fifty thousand feet."

"Two sixty."

"Two hundred and . . ." Palermo paused, awaiting confirmation from mission control. He got it: "Two. Hundred. And. Sixty. Four. Thousand. Feet. . . . Welcome to space!"

The crowd whooped and cheered.

"Still going up," said Palermo. "Apogee: two hundred and seventy-one thousand feet."

Covering his face with both hands, Branson cratered with emotion. His son Sam, standing next to him, placed a hand on his father's back for support. "My dad is not unfamiliar with taking risks when other people are saying, 'It's not possible,'" said Sam, who would ponder that image, the one of his father cupping his hands over his eyes: "That was the definition of a picture telling a thousand words, a thousand sleepless nights." Sam said, "It was just so beautiful to watch everyone's definition of what's possible be blown wide open."

The outline of a single tear streaked the side of Branson's face.

STUCKY WAS FIFTY-ONE miles above the Earth, laughing. He didn't know what else to do. He had been transported to some otherworldly fun house, as sunlight filled the cockpit, lighting up Stucky and Sturckow like stage actors.

Lifting his visor, Stucky looked out the window at the pitch black of space and the majestic blue of the Earth and he could hardly believe his eyes. He grasped for an emotion but could only laugh, incredulously. "Ha, ha. Look outside, man," he said.

He pointed up ahead at the peninsular finger of Baja California. "Check that out," said Stucky. "We can see Mexico."

Baja was a meaningful place for Sturckow; he had spent many long nights driving there to compete in weekend off-road races when he was in college. But Sturckow was not the nostalgic type.

"Yup," he replied, and that was it. If Neil Armstrong "surrendered words about as happily as a hound allowed meat to be pulled out of its teeth," as Mailer put it, Sturckow came from the same astronaut litter. Spellbinding panoramas of Earth filled his windows. "Million-dollar view," said Sturckow. Then he was back to work, consulting his laminated checklist. "Post-space checks complete."

SpaceShipTwo was still inverted with the feather up. Stucky fired an RCS thruster to roll the ship upright and then adjusted the RCS knob to correct their attitude for reentry.

"Spaceship reentering directly overhead the field," Stucky radioed to mission control.

The paper snowflake was floating behind him, but Stucky couldn't see it because he was winched in his seat. He was not one to squander an opportunity, however, so, in order to witness the wonders of microgravity, he removed one glove and let it twirl in the air before snatching the glove and sliding it back on.

"Trim set, still zero. Everything looks good," said Stucky.

"IIP bearing in range," said Sturckow.

"Already called it."

"Excellent," said Sturckow, lifting his hand to block the sun. "Next is twenty KEAS," a reference to their knots equivalent airspeed.

Stucky had one more trick. He tore his friend Jack Fischer's NASA name patch—2FISH—from the Velcro strip on his left forearm and held it up for the GoPro. "Okay, 2FISH. This is for you," said Stucky.

Now they could begin their descent.

"Coming down," said Stucky, as the black sky turned blue and the snowflake fell to the floor and the dense atmosphere buffeted the sides of the ship. They gathered speed. A sonic boom thudded as they broke the sound barrier again.

Sturckow kept an eye on the altimeter. "Fifty-five thousand. Standby." He retracted and locked the feather: "Good lock."

They were almost home. The airport was visible down below. Per the test card, Stucky performed a single roll above the airfield before handing the controls over to Sturckow, who landed the ship.

Stucky unbuckled. He ducked down and out of the hatch, stepping into the crisp California air. To think that ten minutes ago he was weightless, marveling at the overview effect! He puffed his chest as he and Sturckow walked across the runway. They were greeted by enthusiastic applause, a true hero's welcome.

Stucky broke away and jogged over to where Agin was standing along the fence in a fur-lined parka. She was a tough, stoic woman, but had been sobbing throughout the flight—hoping, praying that he would come back. He held her. Of course he still had her ring, he said.

A woman passed her baby over the fence to Stucky. He posed for a picture with the child.

Stucky climbed onstage and took the mic. He was out of breath. "That seems like a long time coming, for some of us, decades," he said, bowing to Branson.

He went on. "There were a lot of firsts here today. But there were also some seconds. There's been two people that have flown multiple winged spaceships to space and back. One was Joe Engle, in the X-15 and Space Shuttle. And now there's C.J. Sturckow, who flew Space Shuttle and then landed SpaceShipTwo." He turned to Sturckow: "So why don't you tell us about it."

Ever the hound, Sturckow said, "Great flight. Can't wait to do it again." That was all.

Then Branson took the stage. "Whoa! Who shed a tear here?" he asked. What a moment of joy, catharsis, and relief for him. He reached into his pocket for a folded piece of paper with some remarks he had prepared in advance, still the clumsy speaker who stuttered and got nervous onstage. "I even forget my lines sometimes," he would say.

Branson unfolded his notes. "How on Earth do I describe the feeling? How on Earth do any of us describe the feeling? Joy? Definitely. Relief? Emphatically. Exhilaration? Absolutely," he said. "People have literally put their lives on the line to get us here. This day is as much for them as it is for all of us."

He finished saying what he had to say and was turning to leave the stage when Stucky called him back: "No, no. You're still here." Stucky had earlier given Branson one of the Earth stress balls, which Branson said he would "treasure forever." But Stucky had a final piece of business. Stucky handed the small black box with Parrish's engagement ring to Branson. Then Stucky motioned to Parrish, who singled his girlfriend out in the crowd.

"Veronica, if you could please come up here," said Parrish. She did. Parrish got down on one knee. Branson handed Parrish the ring. When she said yes, Stucky popped the cork on a magnum and sprayed the couple with champagne.

As Parrish came off the stage another engineer congratulated him and said, "Way to set the bar too fucking high, bro."

THE CROWD DISPERSED when the ceremony ended. Branson, who burned easily in the sun, found a sliver of shade behind a trailer. He was followed by a gaggle of reporters. Moses hung near the back to hear what Branson might say. Branson declared that SpaceShipTwo could well complete its flight-test program within three months.

Moses knew this was highly unlikely but kept quiet; he wasn't there to correct the boss. He felt a tap on his shoulder and turned to see a young woman thrusting a tube of sunscreen at him and asking him to pass it to Branson. Moses furrowed his brow. He wasn't some errand boy. "You do it," he said, glaring at the tube.

Agin was waiting for Stucky when he came off the stage. He fished the ring from his arm pocket and hugged her again.

Dillon stepped up and said, "Give me a hug." They embraced.

I stood behind Dillon, trying to clarify my role. Was I standing there as a reporter? Or as a friend? We all live by some code or another. But we, as reporters, could be especially sanctimonious about ours—burying our biases under the soapboxes we stand upon, preaching the virtues of objectivity.

But this story, more than any other, forced me to reckon with that code, to consider what it meant when a source was also a subject and later became a friend; what it meant when a source provided more than just information but inspiration; what it meant when a source influenced *my* life in the process of writing about *theirs*; what it meant to simultaneously maintain professional objectivity and personal awe; and what it meant when one of those outlier cases, the ones infused with awe, involved my own father.

I concluded that maybe compartmentalization was an excuse, that those who spoke of it as a skill used it to justify inattention. You rarely heard the elite aviator, or athlete, or writer discuss their talent for compartmentalizing anything other than flight, or sport, or the written word; they lived for it. *The rest was noise.*

But perhaps it was also a deception, this belief that life could be boxed into neat compartments—family, friends; work, play; source, subject.

I had invested more than four years of my life into Stucky and his. And though the blinking light of my recorder provided some constant reminder of our unspoken roles—he talked, I recorded—the line between what was personal and professional dissolved over time. I felt his pains and triumphs. When we first met and he told me about his estrangement from his children, I imagined my own two sons someday pretending not to know me, writing me out of their lives. It left me crushed.

Yet I recognized that his loss and pain were powerful components of drama, and that the drama we, as nonfiction storytellers, aimed to capture did not come from some abstract place but had to be dredged up from life. How do you distinguish reportage from exploitation? You were supposed to know it when you saw it. If you were lucky you got to witness more than just pain and loss, and got to see triumph and perseverance on the other side.

While Stucky and Dillon held each other I was contemplating what I would do when they let go, whether it would be awkward to hug Stucky or more awkward not to. How did we characterize our relationship? Friends and family hugged. Writers and subjects shook hands.

I extended my hand, but the gesture felt sterile. I reached around with my other arm for a hug. It felt right.

Suddenly, Stucky remembered something. He went into his calf pocket for one of the Earth stress balls and handed it to Dillon. Some part of me wished that he had one for me. And yet that blinking light reminded me why I was there, and that I had my own my father and my own sons to comfort in moments of sadness and celebrate in moments of triumph.

If there was one thing Stucky missed in that moment, that was it— that his father was not there to see him become an astronaut.

STUCKY AND THE pilots went back to FAITH and drank whiskey in paper cups—a pilots-only affair. When Sturckow, who didn't drink, came into the conference room and saw Jerry Crump, a purchasing manager, sitting at the table, he sniffed, "So, Jerry, you're a pilot now?"

C.J. Sturckow, Richard Branson, and Mark Stucky, in 2018.

Crump supplied the whiskey, said Mackay. He was welcome to stay. Crump finished his cup and left.

Someone suggested Denny's, and when they arrived the staff pushed some tables together to accommodate the large group. Stucky showed up a few minutes late with Agin and Dillon; when they entered the restaurant some of the diners cheered.

Stucky, Agin, and Dillon sat at one end of the table.

At the other end, Sturckow and Ericson were bemoaning Ivy League tuition costs. Both had overachieving high schoolers.

Dave Mackay, the Scot, was arguing with Masucci, a Trump supporter, about Brexit.

Stucky and Agin split a hamburger. He read through his messages between bites.

Alsbury's widow, Michelle Saling, wrote, "I saw the contrails from my front yard and heard the sonic boom so I figured it was a big flight day."

"Bet ya are feeling great," wrote Burt Rutan.

Randy Gordon, an Air Force lieutenant colonel, wrote, "Many years ago you told me that you felt terrible at NASA because you were so close, but not a 'part of the astronaut club.' My hunch was that one day you would go on to be more impactful in the world of human spaceflight than any modern day astronaut in the corps. Today you did it."

Someone read aloud a tweet from Vice President Mike Pence: "The 1st crewed flight to launch from US soil in over 7 years!"

When Stucky saw a post on Virgin's official Instagram account that misspelled his name as "Sticky," he said to Mackay, "I know you're behind this."

Nicola Pecile, an Italian test pilot who had joined the company in 2015 and went by the call sign "Stick," lowered his voice an octave and said, *"And today we're going to fly with Stick and Sticky."*

Stucky was in a generous mood, with nothing to prove, and said how "Sticky" reminded him of his childhood. *"Sticky Stucky stuck a stick into the sticky Stucky mud,"* he said, echoing a playground taunt. "I wasn't beaten up *too* bad."

He got up with a bulging manila envelope under one arm and walked to a nearby table, where four young men with military cuts were ordering breakfast. They put down their menus.

These men were not strangers but pilots from Stucky's former squadron in Nevada; one of them had emailed Stucky earlier in the week to say that he and some colleagues were going to be at Edwards and were wondering whether they could get a tour of FAITH. Stucky told them it was a busy week but if they were still around on Thursday morning at eight a.m., they should look up: they might see something special.

"They got the hint," said Stucky.

After Stucky landed SpaceShipTwo, he invited the guys to meet him at Denny's. He now put the envelope on the table and asked them if they could return the LUCKY T-shirt to the bar when they got back to the base.

A few minutes later, Stucky rejoined Agin and Dillon at the table. Dillon asked how he felt.

Honestly? A swirl, between the human instinct to savor a moment of success and the professional one to carry on and proceed clinically down his checklist. Stucky asked Dillon if he'd told him about the interview he had during his third NASA screening, when the psychologist wished him luck and said that if he got picked she hoped it would be everything he had made it out to be. Her words had always been a bit cryptic and open to interpretation, but Stucky had tried to live by them over the years. He kept them in mind, to help cope with what had always felt like defeat. He never imagined that becoming an astronaut would be anything other than what he had made it out to be.

But now? "I guess it was kind of all I hoped it would be," he said. Though, frankly, his life had not suddenly changed. He was still eating at Denny's.

He checked his watch. They were due back at the hangar in fifteen minutes for the postflight review. "Fifteen minutes till Wapner," he told the table, quoting Dustin Hoffman's character in *Rain Man*. The Space-ShipTwo simulator was only as good as their models, and their models were only as good as their data, and they had reams of new data to feed into the computer—you know, said Stucky, "to see if there's anything more we can learn."

He was already thinking about the next mission.

BACK AT FAITH, after the review, Stucky gave his Air Force friends a tour. He showed them out, and then he wandered around the hangar by himself for a while. Mike Moses observed from a distance. "I mean, he just never left. I don't think he really wanted the day to end," Moses said later.

At one point Stucky went up onto the balcony to look down on the spaceship. He was leaning on the railing when Moses approached. "So?" Moses asked. "What's it like to be an astronaut?"

"I don't know. No different."

"Bullshit. It changes everybody," said Moses, knowingly. Some in subtle ways. Others more egregiously. Buzz Aldrin spent years after his moonwalk in a drunken depression, selling Cadillacs in Beverly Hills.

"I'm already an asshole," said Stucky. Besides, he couldn't afford to get ahead of himself with so much left to accomplish. Today had been a big step for him, but a medium one for the company. "I want to fly other people into space, *repeatedly*," he said.

Moses smiled and called it a day. He was exhausted. Stucky was still there when he left.

Stucky got home later that afternoon. He changed into jeans and loafers and went with Agin to the Broken Bit for dinner. They sat at the bar. Another patron noticed Stucky's sweatshirt with the Virgin Galactic logo. The patron said his son worked there and asked Stucky what he did for the company. When Stucky said he had flown SpaceShipTwo that morning, the man bought him a drink.

Stucky invited me to join him and Agin at the bar, but there were no free stools when I arrived. Stucky offered to go back to his house. He said he had better whiskey there anyway.

Their neighborhood pulsed with Christmas lights that threatened to put an epileptic into shock. Inside, Agin lit the gas fireplace. "You didn't crank that up to high, did you?" Stucky asked, popping the cork on a bottle of champagne.

He stepped out and returned with a coffin-shaped box of whiskey. "Cheryl got this for me a couple of years ago," he said. "I've been waiting to drink it ever since."

It was a limited-edition, fourteen-year-old rye from WhistlePig, a boutique distillery. He asked if I wanted mine straight or on the rocks. I

SpaceShipTwo's view from space.

shrugged; I drank whiskey on only the rarest of occasions; my palate was undemanding and frankly undeserving. "We should at least do a shot of this first," he said.

Stucky went into his office to fetch a wooden rack with a dozen shot glasses. Each glass in the rack featured the logo for a unit based in Tonopah or Area 51. He filled two but refused to comment, confirm, or deny any information about the units. We raised our glasses. What a day.

Clink.

I downed mine in one gulp.

"Holy shit," said Stucky, eyeing my empty glass. I had, in my haste to impress, apparently misunderstood: we were supposed to *sip* a shot glass of expensive whiskey, not *shoot* it in one go. I felt embarrassed and apologized.

The next morning I felt even worse when I saw the bottle online for $600.

31.

THE SPACE MIRROR

A FTER THE SPACE FLIGHT, Stucky became a minor sensation. Backstage passes for the Rolling Stones. A guest of honor invitation to Yuri's Night, the annual space geek gathering named for Yuri Gagarin. Aerion, a budding supersonic aircraft company, offered Stucky a job as their vice president and chief test pilot.

He attended fancy parties in L.A. Hung out with Harrison Ford. He went to a dinner in Beverly Hills for the Living Legends of Aviation with Ford, John Travolta, Buzz Aldrin, Jeff Bezos, and Kenny G. During Kenny G's acceptance speech as a "living legend," the cool jazz saxophonist compared flying ("that freedom!") to playing a solo and made a boorish joke about how his career proved that if you blew something well for forty years it would reward you.

Other perks rolled in. When reps from Stucky's favorite coffee company, Death Wish Coffee, heard of his devotion, they showed up at his door with free beans and a thermos and asked for an interview. "I don't have a death wish, although I love Death Wish Coffee," Stucky said in a clip posted on the company's website. If he had a death wish, he went on, "I'm obviously rather poor at it because I've been around for a few decades doing dangerous stuff."

He flew to Kansas later that year to deliver his former high school's commencement address. He told the graduating class to keep an open mind. For years, he said, "I held the strong belief that there was only one proper way to load a roll of toilet paper. . . . My daughter's cats proved me wrong." He also had something profound to say. He talked about growing up half–Puerto Rican and lower-middle-class, and staying focused on the big stuff. "Some things may be easier if you're born into wealth but there are too many examples of silver-spoon children who have accomplished nothing," he said.

"The real impoverished are those whose minds and willpower are insufficient to commit to chasing their dreams."

MOSES FELT SOME temporary relief: the space monkey was off their back. It was a big morale boost for his team. "This mythical thing is now behind them," he said; now they could focus on building a company that lived up to its advertising.

He tried steering his team's mood from postflight euphoria into preflight preparation, but they took cues from him about what was coming up, and he was taking cues from London about what was coming up, and nobody quite knew what to say. Moses didn't make the strategy. He was an operator. An executor. Give him a problem and he would solve it. He just wanted someone to tell him what problem they wanted him to solve.

"At NASA we called them 'figures of merit,'" said Moses. Definable, measurable metrics—they were critical for any program. He despaired over the absence of such metrics at Virgin. Sometimes it felt like a miracle that they got anything done. The company had remarkable pilots, and plenty of smart engineers with good ideas, but sometimes their success seemed to occur in spite of the company's leadership rather than because of it: Branson spoke boldly without always knowing the facts; Whitesides knew the facts but was often too timid to act.

"All speed, no vector," Todd Ericson would say.

Moses wanted someone to tell him what he should prioritize: Branson flying on the fiftieth anniversary of Apollo 11? Or getting a paying customer on board as soon as possible? Reaching one hundred kilometers? Or flying every two weeks? "Pick your figures of merit," he said. "We can't do all of them, right? I mean, I can, but you've got to give me a lot more money and a lot more time."

He was doubly uneasy about the company getting ahead of itself. He feared that the team in London didn't appreciate the amount of work left to do—the vastness of the chasm between completing a test flight and certifying a spaceline. They were thinking in weeks when they should be thinking in months or even years.

In January 2019, Moses made some PowerPoint slides. The slides reflected his plan for Virgin to conduct at least another six rocket-powered test flights. He sent the slides to his marketing and PR colleagues in London; Branson was about to go on CBS to promote Virgin's new partnership with Under Armour, and Moses wanted to ensure that Branson was informed. During the segment, however, Branson was asked about Virgin's schedule: What was the magic number of test flights before Branson could fly? Three, he said.

Moses groaned when he saw the clip.

Moses was facing some internal opposition, too. This was more personal. He and Whitesides had decided to put a passenger on the next test flight so they could begin gathering human feedback on the cabin dynamics. That information could be used to tailor the customer experience: How severe would the g-forces feel? When should they unbuckle? Would floating passengers disrupt the pilots?

The pilots couldn't run the experiments: they were strapped in, and already far too busy. They needed someone who could pretend to be a tourist, who could absorb an extraordinary experience with everyday eyes, yet someone with training and whose job description justified the risk.

Beth Moses emerged as the leading choice.

When she worked at NASA, she'd logged hundreds of hours in the agency's "reduced gravity" research jet, better known as the "vomit comet," which simulates weightlessness by flying extended free falls, thus enabling engineers to conduct microgravity experiments. "Her job was to run test cards—float around, run this experiment, evaluate that, write down these evaluation criteria," said Moses.

And now, at Virgin, she was in charge of training future astronauts and outfitting the SpaceShipTwo interior. It was her job to imagine the passenger experience, to picture what Virgin's space tourists could, and should, do at which moments in the flight. Where should the handholds be placed? How high should the seats be in relation to the windows? What did the g's feel like in the back, behind the cockpit?

Beth was the right woman for the job. Furthermore, her going would be a statement, evidence of Virgin's desire to "democratize space." Their first "tourist" would not be another tycoon or male test pilot, but a female engineer.

Still, not everyone was sold.

Ericson strongly objected. He thought Moses presiding over a flight with Beth on board was a horrible idea. How could Moses dispassionately manage the mission with his wife on the ship? Ericson thought Moses should at least recuse himself, and that his failure to do so reflected what Ericson, the VP of safety, considered systemic leadership failures—Moses was conflicted; Whitesides was indecisive. How both men still had their jobs after so many years behind schedule was a miracle, thought Ericson: "In the Air Force we fired generals for a lot less."

Ericson and Moses were at loggerheads. Ericson hoped Whitesides would try to arbitrate, but Whitesides was averse to confrontation. "His indecision was a decision," said Ericson. On a conference call, Whitesides let Ericson raise his concerns directly with the board. Ericson argued for putting Nicola Pecile, the test pilot, in the back, and said appointing Beth would set a "terrible precedent." "Just from a good order and discipline standpoint, what are you communicating to the men and women in this company?" he asked.

But Ericson realized by the end of the call that he had lost. The flight plan would go ahead: Beth was going to fly. And now Ericson and Moses were at an impasse.

Still, Moses did not disagree with everything Ericson had to say. He often found Whitesides's indecisiveness to be a problem, too. "It's hard to make decisions around here sometimes," said Moses. He corrected himself: "Sorry. It's hard to get *other people* to make decisions sometimes. I don't have too much of a problem."

What infuriated Moses was hearing Ericson impugn his, and Beth's, professionalism. "This sounds unbelievably cold, but I don't value her life any more than I value anybody else's life in this company," said Moses. "I clearly personally value her. But I would not put Dave Mackay at some higher risk than I would her. Or anybody else. If I have any hesitations about putting my wife in the ship then nobody should be flying," he said. "She knows these risks just as much as I do. And if she's comfortable then I'm comfortable. But we both have definitely chosen not to include our kids in this. They don't get a vote."

He was constantly checking himself, "to make sure I'm not being too clinical about this." But he asked: "What if Beth got hit by a drunk driver coming home from work one night? How would I handle that with the kids? It doesn't make it any easier if she gets killed in a spaceship accident, versus in a drunk driving accident. At least one would be for a reason."

He went on. "Personal life loss aside, an accident would impact the lives of the eight hundred people who work here: we would go out of business. And the commercial space industry could die," he said. "I mean, I have plenty of pressure. Adding my wife to the mix is a drop in the bucket of the other responsibilities I feel."

STUCKY CALLED HIS old friend, Jack Fischer, to book dinner: Stucky was ready to settle his debt.

They decided on a weekday evening at the end of January. Fischer had plans to be at Edwards that week. After work they could meet at Domingo's, just like old times. Word got out. Others asked to come. They soon had reserved half the restaurant.

Stucky and Agin arrived early. Stucky hadn't been there in years, but Domingo's looked the same as ever—pleather booths, test pilot memorabilia covering the wood-paneled walls. Stucky smiled at the silly photos of astronaut crews wearing sombreros, fake mustaches, and burlap ponchos.

Fischer came in with a handful of senior military officers. When he came back from space he had options. Peggy Whitson, a fellow astronaut, was about to retire and go on a lucrative speaking tour. Fischer could have done that, or he could have gotten in line to go back to space. Instead, he left NASA and returned to the Air Force, to help set up the Space Force. Fischer was a warfighter after all, and space travel, despite its comradely and humanistic overtones, had always been a quest for military dominance. "Science happens not because governments care about science. Science happens because governments care about other things that need the science," said astrophysicist Neil deGrasse Tyson.

A waiter took drink orders.

Stucky got a margarita. Fischer asked for the same, but he wanted the big one, the one they served in the stemmed fishbowl glass. Fischer had fond memories of this place. When he was coming up it had been *the* meeting spot for testers.

Domingo Gutierrez, the owner and the town's honorary mayor, made the rules. "At one point when we were testing the Raptor"—the F-22—"he wouldn't let you in if you didn't have a SAP clearance," said Fischer. SAP stood for special-access program, the peak of the top secret pyramid.

Fischer recounted the time he and his wife drove their RV off base, parked at Domingo's, drank far too many margaritas, slept them off in the RV, and then drove back onto the base the next morning. He considered himself an authority on Domingo's. When the general with whom Fischer was traveling asked about the seafood, Fischer suggested the chicken: "We're in the middle of the desert, sir."

They were drinking and laughing and sharing fond memories of acquaintances who had died in the line of duty when Steve Rainey, a gravelly-voiced F-22 test pilot, tapped a knife against his glass. Rainey stood up. He was the one who had provoked the bet, and so he should share the backstory.

"Once upon a time, there was this guy named '2Fish' Fischer, better known as '4Fish.' And he wanted to be an astronaut and somehow he made it. We were at SETP and I guess I threw down a bet: it was based on

Jack "2Fish" Fischer and Mark Stucky at Domingo's, in 2019.

NASA's performance at the time, and Forger was with SpaceShipTwo, and I said, 'Forger, buddy, you're going to beat 2Fish to space.'"

And yet, he said, waving his hand at Fischer, "You won."

Fischer stood—a solid man with a disarming smile. Yes, he said, he had technically won the bet. But, "Forger won by an epic margin of classy awesome."

"The Chuck Yeager of the twenty-first century!" Rainey said, of Stucky.

Stucky shrank a little in his seat, proud but also a bit embarrassed to be feted in such esteemed company; Rainey had more hours in the F-22 than almost anyone, and Stucky considered Fischer one of the finest test pilots of his generation. Dissembling, Stucky remarked that he had been in space for about five minutes, whereas Fischer lived there for four months.

"It took me eight minutes just to get to orbit!" said Fischer.

But Fischer and everyone else in that room knew the truth—how not all flight-test minutes were created equal, how one minute at the controls of an analog rocket ship could say more about a pilot than thousands of hours in another vehicle, how the margin of error on SpaceShipTwo was nearly invisible, and just how extraordinary Stucky's experience and successes had been on that ship, and others.

Stucky stood beside Fischer with his margarita glass aloft.

"Test Gods!" Rainey hollered.

Stucky smiled, undeniably proud. His and Fischer's wager had started innocently enough but morphed over the years into something far more significant, with Alsbury's death and all of the obstacles Stucky would overcome. It meant much more than a plate of enchiladas. It was a tribute to the brotherhood to which Stucky, Fischer, Rainey, and everyone else in that room belonged.

Stucky gave Fischer a still-frame image that Stucky had made from the GoPro footage. It showed Fischer's name tag floating inside the Space-ShipTwo cockpit.

Fischer raised his goblet. "To the future of space flight!" he said.

Hear, hear.

"And to fools like you and C.J. for breaking down those barriers for us!"

STUCKY AND FISCHER met up the following night in the Virgin parking lot. Stucky knew where they were going, so Fischer left his car and rode with Stucky. They drove west over the mountains, into Tehachapi, and into a hilly neighborhood with dirt yards and split-rail fences. Stucky parked outside a one-story home with a basketball hoop in the driveway.

Michelle Saling met them at the door.

She welcomed them inside and asked if they were hungry and offered to make them up plates of barbecue but Stucky suggested they eat afterward. He and Fischer had a few things to give Saling and her kids first.

They sat on opposite couches in the living room. Liam and Ainsley had grown a foot since Stucky remembered. If only Alsbury could see them now.

Fischer introduced himself to Liam and Ainsley. He didn't know their father, he said, but he considered him a kindred soul.

Fischer took out a pair of white glove liners. He had worn the pair in space. Space gloves were sacrosanct; Burt Rutan's friend supposedly refused to even bring his to the local STEM school for fear of losing them.

But Fischer felt differently. He told Liam and Ainsley where his gloves had been and what they had touched; how, when he was on the ISS, the multiplexer—an external computer—had fuzzed out, so he had to suit up and go fix it; how these were the gloves he wore under his EVA gloves— the ones with heaters in the fingertips to protect against extreme cold; and how he now wanted Liam and Ainsley to have them.

"Everyone who is a space explorer takes a big risk because we think it's worth it," said Fischer. "You honor that brotherhood any way you can."

Fischer turned to Saling; he had something for her, too: a matted photo collage from his mission. One was shot through a porthole, with the Earth's blue-and-white swirls in the background, and Alsbury's name tag in the foreground. Fischer included a note: "A REMINDER OF THE AMAZING THINGS HUMANITY IS CAPABLE OF WHEN WE WORK TOGETHER AS 'ONE TEAM.' DARE TO DREAM, 2FISH."

Stucky gave Saling his own collage. His included one shot of Space-ShipTwo's plume set against the black sky, and another with Alsbury's name tag framed against the bright blue Earth. A gold placard at the bottom read: FLOWN ON SPACESHIPTWO UNITY'S FIRST SPACEFLIGHT DEC. 13, 2018.

Stucky asked Alsbury's kids if they wanted to watch the video from his flight. They did. Saling stood behind them. She paced as the footage began. When the countdown commenced she asked Stucky to hit pause so she could ask Ainsley and Liam if they were definitely sure they wanted to keep watching. They were.

Later, Stucky told them how lucky they were to call Alsbury their dad. Stucky said he knew there were a lot of articles out there describing him as their father's best friend but that he wasn't alone. That was what made their dad so special, he said: "A lot of us considered him our best friend."

NOT LONG AFTER, Saling flew to Florida with her kids. Family and friends traveled in from around the country to support her and them, to gather one sunny afternoon at the foot of a forty-two-foot-tall black granite wall on Merritt Island, near Cape Canaveral—the Space Mirror Memorial.

NASA unveiled the memorial in 1991. Fifteen white doves were released at the ceremony. The doves flew up and over the surrounding marshes and mangroves. Each bird represented one of the "pioneers who led America into space," but whose names would now sadly appear on that wall—including three from Apollo 1; seven from the *Challenger*; an X-15 pilot; and my father's friend, Sonny Carter, who had been planning to participate in the dedication ceremony before his plane crashed and he became one of the dedicated.

After the SpaceShipTwo accident, C.J. Sturckow told the Astronaut Memorial Foundation that Mike Alsbury belonged on that wall. The committee did not agree. They asked why they should add someone to a memorial reserved for astronauts who wasn't even an astronaut. Sturckow pointed out that four of the men on that wall had died in training accidents, prior to their first space mission, and therefore technically were not astronauts, either.

The committee didn't budge.

Plus, committee members asserted, Alsbury flew for a company and not for the country. "I'm still not ready to accept the commercial pilots yet," said Lowell Grissom, the brother of Gus, who died in the 1967 Apollo 1 accident. "I just don't quite see that as in step with what's already on the Mirror."

But Sturckow persisted. He went back to the committee after he and Stucky completed their space mission.

"Times have changed," said Sturckow. Any pilot strapping into Space-ShipTwo required "a hell of lot of bravery," he said—on par with any NASA mission. And Alsbury had been a critical member of the SpaceShipTwo program from inception—first glide flight, first rocket-powered flight. He belonged on that wall.

This time they agreed.

Sturckow flew in, as did Stucky and Agin, Alsbury's parents and sister, and several of his former Scaled colleagues. At an indoor ceremony, speakers talked about purpose and sacrifice. "Great gain often comes at the price of great pain," said one.

Sturckow got up and spoke. He did not have a personal relationship with Alsbury, he confessed, but he had tremendous respect for Alsbury's experimental flight-test achievements. "First flights and envelope expansion missions are a *really* big deal," Sturckow said, in his dry, dispassionate manner. And once he and Stucky crested fifty miles above the Earth, there was no doubt: "Mike's work was vindicated."

Outside, a procession moved to the Space Mirror. Hawks circled on thermals overhead in the big, blue cloudless sky. A bagpiper played "Scotland the Brave," and once everyone was standing in the shadow of the wall a black drape fell and revealed the newest name carved into the granite: MICHAEL T ALSBURY. Saling laid a wreath.

The grumble of a jet engine sounded in the distance. An F-104 appeared. It flew low and fast over the Space Mirror, made a second roaring pass, and then, on the third, nosed up with its afterburner blazing. It climbed steep and sharp and fast before disappearing in the infinite blue.

Saling squinted, blocking the sun with one hand. She had felt such despair, uncertainty, and loss all these years. Abandoned. Alone.

The other names on that wall had been wrapped in the flag. State funerals and twenty-one-gun salutes. Grief salved by patriotic comforts.

But Saling had been too desperate and afraid to forget about Alsbury, for even a second. How could she forge ahead when it was her duty to remember him? The sight of his name chiseled in granite lifted that burden: everyone could remember him now.

Suddenly, there was a kind of logic to his death. It was no longer about his actions in the cockpit that morning. Or the children he left Saling to raise on her own. Alsbury was now one of twenty-five noble men and women killed pursuing a dream, trying to slip the surly bonds of Earth.

This stark, shiny wall served as a reminder of that. Now, when people came and saw Alsbury's name and paid their respects, they would do so not with heads bowed but with eyes looking up, way up—into that clear Florida sky that stretched on and on, before finally turning black.

32.

WINGS

ON FEBRUARY 6, 2019, a private jet whisked Stucky to Washington. Virgin had chartered a Gulfstream to take him, along with Sturckow, Moses, and about a dozen engineers to the capital for a ceremony in their honor.

They were traveling solo. This was a business trip. But Stucky insisted on inviting Agin: she had stood by his side for the past ten years while he flew the spaceship and he thought she deserved to be there now, by his side, when he received his commercial astronaut wings.

Upon arriving in DC, they checked into a hotel near the Department of Transportation headquarters, in a redeveloped neighborhood on the waterfront with sapling-lined sidewalks. Administratively, the FAA belonged to the department, and the head of the department, Elaine Chao, the wife of Senator Mitch McConnell, was likely eager for some good press. She was going to preside over the ceremony.

Branson flew into town.

I picked up my parents on the way to the department. My mom didn't know Stucky, and it was forty years since my father had seen him. But Sturckow was a dear family friend who used to attend rowdy barbecues in our backyard. My mom, an irrepressible optimist, couldn't square her impression of Sturckow—his polite, farm boy manner and gap-toothed grin—with the intense, oftentimes inflexible, side of him I witnessed in

Mojave. My father was not surprised; Sturckow didn't earn his call sign, "C.J.," short for "Caustic Junior," because he was a ray of sunshine.

We hustled up the steps and walked through a metal detector. I spotted Stucky and Sturckow almost immediately. They were standing offstage in their flight suits, backs against the wall.

Sturckow looked annoyed.

Chao's deputy had just come over and asked Sturckow how it felt to go to space for the first time. Stucky interrupted the deputy and informed him that it was actually Sturckow's *fifth* time. Away the deputy slunk.

But when Sturckow saw my parents his face lit up. My mom asked about his wife and kids, and he took out his phone to show her pictures and bragged in his modest way about his wife's work at a local middle school. It was remarkable to see the alter ego that emerged.

I had witnessed something similar with my father; he seemed to shed decades of age whenever he got around old fighter pilot friends. His posture, his demeanor, his diction. Everything changed, like he was back in the ready room—mustachioed, smoking cigarettes. A relapse of glory days. Sturckow told my dad that Stucky tried his best to keep that vibe alive. "Forger still treats me like I'm a lieutenant," he said.

Stucky and Sturckow followed Branson and Chao onto the stage. All rose for the national anthem. Stucky felt unexpectedly moved by the performance. It surprised him, because although he had been prepared to die for his country, he was still his father's son and was never one for patriotic pageantry. He realized up there on center stage, however, that this was the first time someone had sung the national anthem in his honor. It made him proud. He knew his father would have been, too.

Chao's deputy gave some clumsy introductory remarks. Commercial space was "an industry that sees no earthly bounds," he said, punning like a traveling vacuum salesman. "If you're looking for the Next Big Thing, commercial space just might be it!" Finally, he said, "Let us *launch* today's program."

Then Chao went up. She stumbled over Stucky and Sturckow's names and tried to blame it on her script; it wasn't her fault, she said: "They were introduced to me as Forger and C.J." She directed her comments to a middle-school class near the front, addressing them like kindergartners. "Our two pilots served in the United States Marines. They're *tough*." She pointed at Sturckow. "That's a real-life NASA astronaut, kids. That's what they look like. See? Tough. Terrific. He flew on four Space Shuttle missions to the International Space Station. He's left Earth!"

Stucky looked over at Sturckow who was staring straight ahead, steely as ever. He knew Sturckow hated this kind of thing, which made it that much more enjoyable.

"C.J. has got to be dying right now," my mom whispered.

When Chao finished, Stucky and Sturckow rose for the pinning of their wings. Chao asked Stucky if she could unzip his flight suit. He obliged. Then he, Sturckow, and Branson posed for pictures. Stucky reached behind and put bunny ears on Sturckow. A photographer captured it.

The next day, the *Washington Post* published an article about the ceremony. It was spread across four columns and accompanied by a photo of Sturckow, with bunny ears.

AFTER THE CEREMONY I hurried across town to take my sons out of school. They looked surprised to see me when they shuffled into the office; they weren't expecting to get out early. I asked if they wanted to go to the National Air and Space Museum to see the SpaceShipTwo rocket.

Yes!

Ages four and seven, the boys were obsessed with rockets and airplanes. Small wonder. Sleepovers at my parents' house typically devolved into extended YouTube sessions involving jets and motorcycles. My dad showed them how to fold their paper airplanes to achieve maximum range and stability, and he took Oscar, my older son, flying in a rented taildragger. (Bohan, my younger son, was too small for the harness.) Later, when assigned to write an essay about bravery and fear, Oscar wrote, "The bravest person I know is probably Poobah because he flies really fast planes and drives really fast motorcycles and he is not scared at all."

The only version of me that Oscar and Bohan knew was the one immersed in this book. Excitement is contagious. They preferred *The Right Stuff* over Pixar, adventure over animation. Oscar would celebrate his eighth birthday by going to see *Apollo 11* at the IMAX theater in the Air and Space Museum. We built model rockets at the kitchen table.

They were endlessly interested in learning more about SpaceShipTwo. Being kids, they wanted to know all about the crash. Their interest was genuine. But I also suspected it was an excuse, a pretext for them to learn more about me. A way for us to talk about more than school. Or what we were having for dinner.

I suspected as much because I was doing the same thing—writing this book as an excuse to learn more about my dad.

So I answered their questions: about the scar on my left cheek, and whether I remembered being sideswept by the eighteen-wheeler, or being rushed to the hospital in a helicopter (I didn't); about wild boar hunting with my dad; about the night I spent in jail; about doing journalism, and whether it was scary interviewing terrorists; and always, because children have morbid imaginations, whether I thought I was going to die.

I tried to tell them the truth. About the power and temptation of fear, and how even the brave got scared, but how our moments of greatest fear were often the most memorable.

I don't remember having those conversations with my dad; fear was an admission of weakness, and weakness was one of those words that you spat rather than spoke. If my dad talked about fear and risk, he did so with the cold distance of a Hemingway character, and not one of the flawed ones, because no one wanted to grow up to be like Francis Macomber, a caricature of cowardice.

But at times I feared that this was precisely who I had become—fronting as brave, but racked with doubt and cowardice. I remembered being thirteen in the parking lot outside a Dairy Queen when a handful of older teenagers were cursing loudly, and my dad, who didn't appreciate them using that language in front of my mom and my brother and me, went and told them to watch their mouths. Now, I wondered: Would I do the same? Or would I pretend I couldn't hear them while telling my wife and kids to mind their own business?

That was what scared me most. What kind of man was I? What kind of father did I want to become?

We live enveloped in whale lines.

OSCAR AND BOHAN hurried home and dropped their backpacks. My wife soon arrived and my parents pulled up; we squeezed into their car for the short trip to the museum. There, we followed signs to an exhibit near the back.

Branson was standing at the mic, praising what he called his favorite museum in the world. "If you notice I've gone AWOL later on, it may well be that I've found a place in this wonderful building where no one will find me before locking up for the evening," he said. He remarked on how this incredible museum had become a kind of showcase for Burt Rutan. "Without Burt Rutan, the rest would not have been history," he said.

Briefly, Branson turned solemn, noting the "great sacrifice" that many had paid leading up to this day: "People have literally put their lives on the line to get us here, and all our important days are as much for them as they are for us."

Stucky appreciated the sentiment but wondered why it seemed so hard for Branson or Whitesides to ever mention Mike Alsbury or the three cold-flow accident victims—Glen May, Todd Ivens, and Eric Black-well—by name.

Then Branson, ever the showman, turned our attention offstage to what looked like a pommel horse draped in a black cloth. He counted down and, like Houdini, yanked off the cloth, revealing the very rocket motor Stucky and Sturckow had flown to space. Branson was donating it to the museum.

Ellen Stofan, the head of the museum, thanked Branson and said that she hoped this would help inspire the next generation of explorers.

Branson agreed. He mentioned his own grandkids and said, "We need to inspire them."

My youngest stood butt-level in the crowd and complained that he couldn't see. When I bent down to hear him better he told me that I should brush my teeth.

Before everyone dispersed, Stucky asked for the microphone. He told a brief story about visiting an old friend a few years back. One of his friend's sons had walked in wearing a yellow T-shirt with the words FUTURE ASTRONAUT TRAINING PROGRAM printed across the chest. Stucky complimented the son on his shirt and asked where he could get one. The dad told his son to give Stucky the shirt.

Stucky didn't wear it for years; doing so would be presumptuous, he felt. But now he'd made it. He was an astronaut. "I don't think I need this shirt anymore," he said, handing it to Branson.

Promptly, Branson removed his leather jacket and went about unbuttoning his white Oxford. "I don't think this has been done at the Smithsonian before," he said, baring his chest. He pulled the yellow T-shirt over his head and thanked Stucky. "I will wear this for four or five months," he said. "Then I will hopefully get one of your badges and I'll be able to pass this on."

A voice chimed over the PA system: "The National Air and Space Museum will close in fifteen minutes."

MY KIDS HAD a week off from school. We flew to Los Angeles and rented an RV.

The RV owner met us at the airport to hand over the keys and demonstrated how to operate the wastewater valve without getting covered in filth. We drove up the coast.

We were preparing to move abroad. My wife had accepted a new job in London. Eagerly, the boys and I would follow: I could write anywhere; London sounded like an adventure to them. Their passports were already full of stamps. But before we packed up to go live in another country, my wife and I wanted the kids to see, and appreciate, more of ours.

Those were dreary and cynical days. Oscar would read the newspaper over my shoulder—articles about war, poverty, corruption, climate change. Bohan asked questions about the pictures: scenes from natural disasters; portraits of scared, desperate, angry people. Always, breaking news that seemed to heighten a sense of our uncertain future. The newspaper was an awful place to look for role models. Heroes were hard to find.

We camped on a hillside overlooking an inlet and went to a butterfly grove the next morning, bending our necks to peer at monarchs in the treetops. Later, we drove up the coast to watch blubberous elephant seals wrestle on the beach.

Then we headed to Mojave.

I had known for some time that a big test flight could happen that week. Yet those flights so rarely stayed on schedule that I just assumed it would slide and that we would miss it.

Remarkably, the date stuck. SpaceShipTwo was ready. The pilots were ready. Beth Moses was ready. Branson arrived. It all came down to whether the winds would hold. "It's a toss-up," Stucky said the night before.

We pulled into town after dark. I had been here so many times—this was my fifteenth visit—that Mojave felt like a second home. But not one of *those* second homes. It was forlorn. Symbols of desperate dilapidation wherever you looked. A town without a pharmacy, because pill junkies would raid the shelves. Locks on the gas pumps to deter fuel thieves. Motels that competed for guests by hanging weather-beaten signs that advertised AAA discounts and HBO. We checked into the Spaceport RV Park, a concrete slab near the airport surrounded by a chain-link fence.

On our way to dinner I showed my family around town.

There's the Denny's, I said.

And there's the doughnut shop.

The airport is over there but you can't see it because it's dark.

We went to a Mexican place for average tacos and sweet *horchatas* that we sipped from giant Styrofoam cups under buzzing fluorescent lights. When we left the taqueria the winds were picking up, and it blew hard all night, rocking the RV and whistling through every crack. We woke up to discover the inevitable: they had canceled the flight because of the wind.

Making the best of it, Stucky invited us to the hangar so Oscar and Bohan could look around. He met us in the lobby near the candy bowl. The receptionist gave us temporary badges. Bohan posed behind the headless cardboard astronaut cutout, next to the one of Branson in a spacesuit. We followed Stucky down the hallway adorned with glamour shots of SpaceShipTwo and WhiteKnightTwo, and the laminated pictograms that illustrated hazmat procedures, through the pair of double doors and into the hangar.

Gobsmacked, my kids turned mute and wide-eyed, like the first time they stepped through a tunnel and saw the reptilian green grass on a floodlit ball field at night. The hangar was brighter and whiter than an operating room. Its starkness accentuated WhiteKnightTwo's albatrossian wingspan and the coiled, compact package of SpaceShipTwo suspended underneath.

Stucky sidestepped a stanchion and waved us through, explaining the air-launch system as we approached the mated vehicles. My kids had seen enough SpaceShipTwo videos to know how it worked, but they were rapt, listening like it was the first time—unblinking, mouths agape.

Stucky asked if they had any questions. Oscar raised his hand. He was tender and handsome, with a gorgeous mess of blond curls that he tried to comb straight because he got embarrassed when his sensei called him "Curly"; I worried that he cared too much about what other people thought, that his looks and temperament were setting him up for child stardom and tabloid disrepute.

Or maybe not, I now thought. His brow furrowed as he peppered Stucky with questions, not once looking back to me for approval. Sharp questions, about how the SpaceShipTwo pilots knew the proper moment to release. Like, What if they accidentally dropped too early? Or during takeoff? And how *did* they release? And why didn't the plume from the rocket motor singe WhiteKnightTwo as the spaceship blasted off? Later, Stucky asked whether I had planted those questions. No, I said truthfully.

Suddenly, Bohan got excited.

Look, he said, pointing across the hangar. Branson was showing around a group of Virgin Atlantic flight attendants. I know that guy, said Bohan: "That's the man who took off his shirt in the museum!"

VIRGIN RESCHEDULED FOR later that week. We left and drove to Joshua Tree.

Stucky had a strange feeling. He rarely got this way when he was flying. More when he wasn't, like the clutch quarterback who can't bear to watch his teammate attempt the winning field goal. He just preferred being in control and feared complacency, but it was hard to check others the way he checked himself. Stucky knew his fellow pilots were the best of the best. That didn't make them, or anyone else, immune from mistakes.

In December 1963, Chuck Yeager had been arguably the finest test pilot alive when he crashed his NF-104A, an experimental rocket ship. The episode would become enshrined in flight-test folklore—the rare moment when even the irrepressible Yeager couldn't save his plane. Tom Wolfe described the flight in *The Right Stuff*, and it was later re-created for the film.

According to conventional wisdom, Yeager had been zooming 100,000 feet up when the airplane stalled, suffered an RCS failure, and began to tumble. Valiantly, Yeager tried to hold on but ultimately had to eject to save his life.

Stucky knew there was more to the story. Over the years he had read up on the incident, so it bothered him to hear the departure blamed on an RCS failure instead of Yeager exceeding the specs of the envelope, insisting (over the objections of the engineers) on taking the plane up before flying the simulator first, pushing the vehicle too far, and losing control. The airplane hadn't failed him; he failed the airplane.

This terrified Stucky—personally and as a company. He valued experience but worried about his seasoned, accomplished colleagues losing their edge with so little left to prove. Stucky almost thrived on self-doubt and inadequacy, constantly asking himself whether he was sufficiently prepared. These doubts were probably excessive, he knew. He couldn't expect the other pilots to spend their weekends perusing old flight manuals. But he could police them on the job. Recently, he had confronted Mackay when he thought Mackay wasn't pushing himself hard enough in the simulator. Hadn't Mackay learned anything from the July flight that went awry?

"This is bullshit," Stucky told Mackay.

Later, when Stucky found an analysis online of Yeager's NF-104A accident, he forwarded the paper to his colleagues—an oblique reminder that bad days could happen to great pilots. "Failure to admit mistakes is a cancer that must be nipped at the bud," Stucky wrote.

On the day before the rescheduled flight, he drove home, wondering if he had asked all the questions he should have asked. He was sure there were some out there that he didn't know to ask—the ones that were neither the known knowns or the known unknowns but the unknown unknowns. Those were the ones "treacherously hidden" beneath the "subtleness of the sea"—the ones you didn't know to look for.

And there were hidden ones, indeed. What neither Stucky nor anyone else knew was that after some technicians had stripped a layer of thermal protection off the h-stabs, they had re-covered the surface with Kapton, a polyimide film commonly used on satellites because of its resilience under extreme temperatures. But in doing so, they failed to notice the holes on the surface of the h-stab. The holes were designed to vent high-pressure air. Now they were sealed with Kapton. No one would notice until it was too late.

ON FEBRUARY TWENTY-SECOND, I woke up at four a.m. and drove three hours straight to Mojave. We arrived just in time to park the RV and hustle to the flight line to see WhiteKnightTwo take off. My kids each grabbed a Danish from the heated tent. There was a stage, but Branson had left to organize a benefit concert in South America; his absence meant less hullabaloo.

The planes ascended, heading north, and disappeared into the periwinkle sky.

We brought a pair of binoculars. Oscar and Bohan took turns aiming them at the sky. I offered $5 to the first one who spotted the contrails but warned them that it would be at least thirty minutes. Every two minutes they asked how much time had gone by, and they fought over whose turn it was with the binoculars. There was money on the line.

I saw it, said Bohan.

Oscar claimed the same.

I crouched down to their height to see, from their vantage, what they were seeing. What they thought they saw was no more than a faint fleck of high cirrus. The drop point was farther west.

Look over there, I said, pointing just above the ridgeline of the Tehachapi Mountains.

We squinted into the sun. It reminded me of being a kid standing on the runway between my mom and my brother awaiting my dad's return from war, our eyes on the horizon, intent on being the first to spot that glimmer in the distance of sunbeams reflecting off his jet.

Combat and space. They were two sides of the same envelope, foreboding environments where elite pilots went to test their mettle, to gauge how they would perform when the stakes were highest. But what I realized then was the superiority of a yet more noble testing ground: fatherhood. It didn't require a weapon or a rocket, and you got your chance every hour for the rest of your life to perform when the stakes were highest—to model, to inspire, to encourage your children to relish the effort required to perform "magic," to remind them that it was "just someone spending more time on something than anyone else might reasonably expect."

A few minutes later, WhiteKnightTwo's twin contrails appeared; the plane itself was but a faint apparition in the distance, twenty miles away, but the vapor trails were distinct.

Oscar and Bohan each insisted they saw the contrails first. I ruled it a tie.

Enrico Palermo, the designated MC, counted down from five and repeated the commands he heard coming from the WhiteKnightTwo cockpit: "Release, release, release."

And then, ignition.

A burst of fire appeared high over the mountains in the distance and for a second you couldn't tell what was happening because the spaceship and the mothership were so close, but then there was a flame shooting out from SpaceShipTwo's tail, the flame Oscar was rightfully worried about.

The mothership banked one way. The cometic plume kept going up.

Oh my God, muttered Oscar.

I was still squatting between him and his brother, watching what they were watching. I was Oscar's age in January 1986 when my second-grade teacher wheeled a TV into our class because she wanted us to see the *Challenger* take off. When the Shuttle blew up we were sitting Indian-style on the floor. I remembered the silence that followed, but I couldn't remember what she said to break the silence, how she explained that horrifying, curlicued explosion to a classroom of seven-year-olds. I prayed now I didn't have to find and share those words.

Instead, I shared only their awe. We craned our necks to see, but it felt appropriate that we were craning them as gestures of humility, astonished by the courage required to push the envelope, to go *beyond*.

I realized things that morning. Like how I wanted to see Stucky and Virgin succeed, not because I cared who won the private space race but because their passion felt pure, an antidote to the cynicism infecting America; the inside of FAITH felt like a bubble, a sanctuary from bitter politics, a place where intellect was prized, stupidity renounced, and dreaming encouraged. Then again, maybe Virgin's ultimate success and fate didn't even matter: the story had already worked its magic on me. I had taken refuge in its hopefulness during an otherwise often hopeless time.

When I sat down to write I could conjure their world, hearing their voices and seeing their faces and losing myself in the purity of their pursuits. I could ignore—you might even say, compartmentalize—the so-called real world.

The rest was noise.

There was a presumption baked into that statement, a belief that our noblest passions justified single-mindedness. But who determined the nobility of those passions? It occurred to me that perhaps the best measure of nobility was the degree of wondrous awe those passions produced in others, whether those passions would, as the inscription read on that red balloon from Alsbury's memorial service, INSPIRE US TO NEW HEIGHTS.

SpaceShipTwo continued going up.

"Mach three-point-oh," said Palermo, as the motor shut down and the spaceship glided the rest of the way into space.

Bohan kept his eyes pointing up and said, "They are *so* high."

BETH MOSES PRESSED her face against the window. Her jaw dropped. The light in the cockpit dimmed. Her mind struggled to sort what she was seeing into familiar categories. It looked unreal but felt all so very real.

She had trained for this moment her whole life. She knew what she had to do and what she was supposed to see, down to each split second. But nothing prepared her for *this*—the palette of "deep, deep black" and ocean blues and green terrains and the snowcapped mountains below. "I thought Earth was wearing her diamonds for us," she said.

Once cleared by Mackay, Beth knew she was safe to unbuckle and move about the cabin. As she did so, her body floated. Her blond ponytail drifted up. She conducted several tests, pulled her way to the front with her white-gloved hands on the handholds, and poked her head between the two pilots.

"You can see forever," said Mackay.

A voice crackled over the radio from mission control. "Welcome to the club, astronauts," said Stucky.

"I like this club," said Mackay. "Welcome to space, Scotland."

Beth checked her watch and saw it was time to return to her seat. As rehearsed, she rebuckled and said over the radio, "At my seat." They began their descent.

A part of her couldn't wait to come down and evangelize, to pour forth with giddiness, to try but know that she could never put into words what she had seen and felt up there. But another part of her would feel some creeping unease about the here and now and the future. "What am I going to do for the rest of my life?" she later asked.

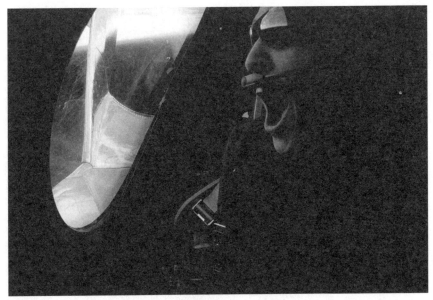

Beth Moses.

They bumped and skidded to a stop.

Planeside, a bagpiper played as Mackay got out. He danced a Scottish jig and hailed "the end of a long journey."

Beth was handed a bouquet of flowers. Her two daughters and her elderly parents and Moses stood just offstage.

Moses had been fighting the flu and a high fever and hadn't slept well in days, and as much as he wanted to hold it together because that's what everyone expected him to do, when he looked at his daughters staring proudly at their mom and he looked at Beth's parents staring proudly at their daughter, his eyes welled and he felt an almost crippling amount of joy and relief. "Emotions unloaded fast," he said.

Beth cradled the bouquet with her white gloves and took the microphone with her free hand. She heaved with excitement.

My sons looked up at her on the stage with awe. She tried to catch her breath but it was all too much to put into words.

That, she said, was "an indescribable ride."

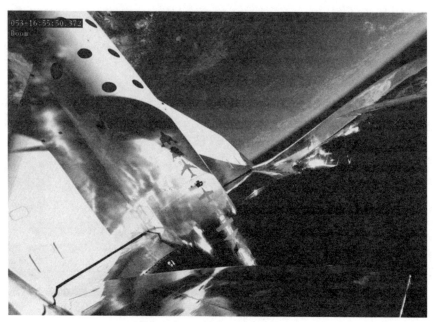

SpaceShipTwo in space.

EPILOGUE

MAGIC

BETH MOSES LEFT the stage amid great optimism and hints of greater things ahead. Virgin had done it, proven the concept, made an astronaut of someone other than a test pilot. What next? "We're in the fiftieth anniversary year of Apollo," said Whitesides, who harbored hopes of making Richard Branson an astronaut on the July anniversary of the maiden moonwalk.

Those hopes did not last long.

Soon after Beth walked offstage, the crew wheeled SpaceShipTwo into the hangar. Stucky and Ericson immediately saw what none of them wanted to see: a wide gap running along the trailing edge of the right h-stab. "It looked like someone ripped the caulking out of a bathtub," said Stucky.

The seal had disbonded on the way up, as the pressure increased with nowhere to vent. When Moses arrived and saw the gap, he felt his stomach sink.

"The structural integrity of that entire stabilizer was compromised," said Ericson. "I don't know how we didn't lose the vehicle and kill three people."

I reimagined Oscar's "Oh my God," said in horror rather than awe.

That night, Mike and Beth Moses had dinner with their kids and Beth's parents. Beth was still giddy; she knew about the delamination but she and Mike did not discuss it at the table.

The next day, Beth's mom and dad drove back to Tucson. When they got home her dad didn't feel right. He woke up the next morning, collapsed from a heart attack, and died.

VIRGIN GROUNDED SPACESHIPTWO. They would not fly it again for over a year. Management tried to downplay the h-stab problem, worried that it might spook customers.

The response worried Ericson. "This should have been a Come-to-Jesus Moment, not the kind of thing you brush under the rug," he said.

Ericson urged Moses and Whitesides to treat the problem as an organizational failure: the maintenance crew had supposedly inspected the ship and verified that it was safe to fly when it demonstrably was not. Ericson despaired that they were treating it as an isolated incident.

(Moses said, "We were nowhere near the h-stab coming off.")

Virgin began redesigning the h-stab and hired a contractor in Wichita to build a new one from scratch, out of metal. Moses expected this to take months.

On April fifteenth, Stucky got called into a meeting with Moses. Moses informed him and Mackay that the company was moving to New Mexico. They were going to have to relocate eventually, Moses said. Might as well do it now, while they were grounded; when the new h-stab was ready they could resume the flight-test program from Spaceport America.

Stucky and Agin flew to New Mexico to look at real estate before the locals got wise and started jacking up home prices. When Stucky got back to California, he went to see an orthopedic surgeon. Stucky knew he had some time before the spaceship would be ready, and he wanted to have an elective surgery done to fix his aching knee. He was put under, and at first the nurse didn't believe Stucky when he said he was an astronaut because he sounded like a lunatic as he talked about his job and his hobbies.

Later, the nurse asked him what he was looking forward to doing with his new knee. Jump off some mountains, said Stucky. He still sounded like a lunatic, but now the nurse knew he was telling the truth.

Stucky recuperated at home in a full-leg compression sleeve. He was intent on avoiding his painkillers so he could keep up on emails—he didn't trust himself on Vicodin—but ultimately succumbed to the pain.

Agin made them hot dogs for lunch on a Foreman grill. She wasn't used to having him around during the day. Each tampered with the thermostat when the other one wasn't looking. Stucky was burning up from the painkillers. His leg felt like a sausage in that compression sleeve. "Did you turn up the heat?" he asked.

She said she was cold.

Stucky said, "I wouldn't want to torture you when I can torture myself," but he asked if Agin could turn it down.

She complied and left to put on a sweatshirt.

ERICSON THOUGHT MOVING to New Mexico was a horrible idea. Now? In the middle of the test program? They would stretch themselves thin and sever their supply lines. It was folly, like Napoleon invading Russia, one of those fateful decisions that MBA students would one day scrutinize and wonder how the company's leadership hadn't known better.

But his misgivings went beyond New Mexico. He had lost confidence in the maintenance team and, by extension, in Moses and Whitesides. It began with their disagreement over Beth. But their response to the h-stab crisis worried him even more. Ericson believed that members of the maintenance team must have been "pencil-whipping" inspections—signing for inspections that were not conducted properly; which might explain why maintainers had not only failed to notice the blocked vents but also missed a baggie of screws taped to the inside of the h-stab. Ericson recommended firing the maintenance director. Moses refused.

Ericson's job gave him a direct line to the board of directors; after the crash, he had briefed them regularly on the investigation, and he maintained good relations. In early May, Ericson got an email from a board member who had heard about his frustrations; they spoke on the phone, and Ericson was invited to brief the board's safety committee a week later, which he did. He told them that failures in the maintenance organization

were making the program unsafe, and if something didn't change someone was going to get killed.

The board hired an outsider, Dennis O'Donoghue, to review the program in search of weaknesses and vulnerabilities. O'Donoghue was a former marine and NASA test pilot. He had spent a decade at Boeing before
retiring in Oregon to grow pinot noir grapes. He was the right guy at the
right time for the job. Boeing's 737 Max was having all kinds of problems—
two recent accidents that killed hundreds of innocent people, grounding
the fleet—and O'Donoghue, who had not conducted the safety review of
that jet, felt somehow responsible; according to one Virgin employee who
spoke with him, O'Donoghue felt confident that had he done the review,
he would have caught the problem and those people would still be alive. He
was convinced that if there was a glaring safety issue on SpaceShipTwo, he
would find it.

O'Donoghue spent weeks in Mojave, asking questions and digging
through files. He interviewed Stucky, Ericson, and others. But regardless
of what O'Donoghue did or didn't find, Ericson knew that his time was
probably up. His disagreements with Moses ran too deep: "Either I needed
to go, or he needed to go." (Stucky considered Ericson's move an act of
betrayal, an ultimately unsuccessful coup cloaked in the fretful language
of safety.)

On June tenth, Ericson told Whitesides that he was stepping down as
VP. Whitesides looked stricken; his vice president of safety was resigning
because he'd lost confidence in the safety regime.

A month later, O'Donoghue completed his inquiry and submitted
his report. Virgin, citing confidentiality agreements, declined numerous
requests to share the report but said that they concluded it was safe to fly.

ONCE THE HOT flashes passed, Stucky started rehabbing at home, using
those thick elastic bands that looked like lobster claw restraints. He recovered quickly. They put a down payment on a home in Las Cruces, New
Mexico.

By August, his knee felt almost as good as new and he was cleared to fly. Since SpaceShipTwo was still in the shop, he flew the Extra and got in the simulator. He surely missed this, but what he missed most was flying with Dillon. Stucky had bought them each a new, high-performance glider and invited Dillon over to leap from some nearby hilltops.

It was not a glorious day. Both felt rusty and a little overwhelmed by the new glider. Stucky had watched Dillon fly for a few minutes before he launched. Stucky had been amazed how quickly Dillon learned. A natural talent, he thought. But now, looking up, he saw Dillon performing a risky landing maneuver, and Stucky, who believed that risks were fine so long as you took them wisely, feared that his son was going to hurt himself.

Stucky got on the walkie-talkie and told Dillon his approach was too aggressive. When Dillon finally got down he admitted to having scared himself up there.

That's normal, said Stucky.

Or at least that's what some people called normal, that fear impulse surging through the almond-shaped cluster of neurons in our brain, dilating our pupils and sending our heart rates through the roof.

But the truth was, Stucky couldn't say for sure: his normal was not normal.

A MONTH LATER, Stucky drove a fancy toilet halfway across the country. He and Agin had been in Istanbul a few years earlier when she encountered a bidet for the first time. She had initially cringed at the contraption, snickering, and Stucky told her the story about the marble bidet in the Baghdad palace that once belonged to Saddam Hussein. By the time they flew home she was converted.

Back in California, Stucky ordered an all-in-one toilet-bidet. It was heavy and was going to require some creative plumbing, and it always felt like they were about to move to New Mexico, so he kept it boxed up in the garage. But now they had a house in Las Cruces with contractors on-site doing renovations.

Send it here and we'll install it, they said.

Stucky didn't even bother getting a quote from UPS; he knew it would cost a fortune to ship. So he lifted with his legs and heaved it into the back of his SUV and drove twelve hours across the desert by himself.

He used the time to think.

Maybe Virgin was going to become a viable business after all, he thought. SpaceShipTwo remained down, but the company's fortunes were looking up. Chamath Palihapitiya, a billionaire former Facebook executive and venture capitalist, came up with a plan to take the company public and had thrown in $100 million of his own money to show his commitment. Virgin was suddenly flush with cash.

Stucky had also been anxious about selling his house. It sat on the San Andreas Fault and he felt like they were living on borrowed time and due for a major quake, but the house sold easily and he even turned a profit.

They booked the moving truck for mid-November.

On October twenty-fourth, Stucky drove into work. It was a Thursday, four days before Whitesides, Moses, Branson, and Palihapitiya were due on the floor of the stock exchange for the big IPO.

Stucky took his lunch break at the local ball field. He parked and removed his paraglider from the trunk. It was sunny, and the winds were manageable, and the forecast didn't indicate anything otherwise. A pleasant Mojave day.

He pushed open a chain-link gate, walked past one of the dugouts and across the infield. The field was in horrible shape—the dirt like old cat litter, the outfield weedy and overgrown—but it suited his needs.

Stucky unpacked his red-blue-and-orange chute and spread it across the parched grass. He had decided after his own disappointing flight that day with Dillon that he ought to take his new paraglider to an open field where the breeze would fill the chute so he could practice yanking on the lines to see how it felt. Paragliders called this "kiting." Think of a bird on a tree branch pumping its wings, or a racer revving his engine in the pit.

Stucky noted the outfield fence and some power lines, but he had plenty of room. He laid out his harness, putting a foot through one leg loop, then the other. A breeze rolled in. He peacocked the chute open and tugged on the lines, and then collapsed the wing again. He felt more comfortable each time; he had to tell Dillon, he was excited to get out and fly this thing for real.

An opportunity came sooner than he hoped.

A strong gust suddenly blew in. It caught Stucky's open chute and began to pull him away, lifting him off the ground. The toes of his sneakers dragged along the dead grass. And then they were dangling in midair.

What can I tell you about this first step that encounters nothing solid?

He first read those words almost fifty years ago in *National Geographic*. They were the words that had drawn him into the sport; he would close his eyes and imagine how it might feel, that sensation of swimming feet. He had imagined the freedom and independence, an ability to chart his own course, a way to temporarily escape the world, to sever those whale lines. Could it really be as magical as the article made it sound, how that first "upward stride causes the jaw to drop and the mind to cease its disciplined churning"?

Indeed, he had discovered: it was everything he hoped it would be.

But now his feet swam and he found nothing romantic about any of it. He got swept up and away, like a theme park balloon. It happened so fast: on the ground one moment; fifteen feet in the air the next. He looked up and saw the power lines ahead. He feared getting snagged and electrocuted if he tried to ride it out.

His mind churned. He was too low to throw his backup chute. And he could tug on the C-risers to collapse the wing, but it was a long fall onto the infield dirt.

He looked up again, considered the high-voltage tangle waiting for him, and decided to take his chances with the ground. If he could land on his feet he would avoid damaging his head or his organs. He hoped his bones were ready.

He fell like a brick onto the hard, unraked infield. He screamed when he hit the ground.

He was alone and it was lunchtime on a school day, so the park was empty. He lay there with his cheek on the dirt and saw the taunting spout of a funnel cloud go skipping by; a dust devil had gotten him once again. He tried to get up but couldn't. Both feet were broken, he feared.

He unzipped his phone and dialed an engineer on his team named Andy Goodman. Stucky had hired Goodman, a mathematician who previously worked at Boeing, over lunch at Denny's.

Goodman picked up the phone.

Groaning, Stucky asked Goodman whether he thought he could sneak out of the hangar without anyone noticing; he had hurt himself and didn't want anyone in the office to find out just yet.

Of course, said Goodman.

Stucky sent him a dropped pin.

Goodman arrived and lifted Stucky off the dirt. Goodman helped Stucky into Stucky's car. Stucky drove himself home. He called Agin on the way and asked her to get ready and to maybe grab a few of the painkillers he had remaining from his knee surgery. She rushed him to the hospital. An X-ray confirmed Stucky's suspicion: a shattered left heel and a broken right tibia.

At midnight, Agin rolled him out of the hospital in a wheelchair.

ON THE FOLLOWING Monday morning, Branson stood on the floor of the New York Stock Exchange in a custom blue space suit, banging the brass bell to open trading as pyrotechnics fired off the monitor posts.

"He's magic," said Jim Cramer, the bald, gesticulating CNBC host. "People want a little bit of magic."

But Cramer was less convinced about the stock. "It's a lottery ticket," he said. Space tourism? Sounded risky. Then again, so were all big technologies before they weren't. And people wanted magic. They wanted to dream. They wanted to believe; maybe Branson knew something special. "I think there are people who think, 'You know what? We don't know what the future looks like. But he does,'" said Cramer.

Mark Patterson, the race car–driving investor and onetime Virgin customer, remained bearish. I had recently gone, with my wife and kids, to watch him race a six-hundred-horsepower Le Mans Prototype on a Grand Prix track northwest of London. (His team finished seventh.) Patterson wished Virgin all the best, but he couldn't see a future for them. How did they expect to ever turn a profit? Patterson had done the math. It didn't add up. "Unless they significantly change their strategy, no matter how shiny and pretty the rocket looks in space, this company is doomed," he said.

But Virgin's viability as a business was separate from the prospects of the commercial space and space tourism industries: those looked solid and promising.

Luke Colby, the pith helmet–wearing propulsion engineer, said, "At the end of the day, I think Virgin Galactic will be remembered for being the first, but not necessarily the most successful. That gives me some sadness. But I'm also hugely proud of what we did."

Colby had recently done some consulting for Blue Origin and SpaceX. Both were bound for success, he thought. "If you want a space vehicle to be fully reusable, for airline-like traffic, it doesn't have to look like an airplane. It just has to function like one. That's the key to making space travel a reality," he said. "And it turns out that the physics really drive you toward a two-stage, vertical-take-off-and-vertical-landing rocket. When you're reentering the atmosphere hypersonic, it's much easier to deal with a cylindrical, tubular rocket coming in tail-first and landing on legs than a winged vehicle like SpaceShipTwo." Colby retained all of his admiration for Virgin's audacity and SpaceShipTwo's aerodynamic novelty. But, he said, "Blue Origin and SpaceX have been driven by physics."

Blue Origin was building a lunar lander at its facility outside of Seattle and an orbital rocket at its Kennedy Space Center complex near Cape Canaveral.

SpaceX, meanwhile, continued to outdo itself: in February 2018, the company launched the biggest rocket since Saturn V carried Americans to the moon, putting a Tesla Roadster into orbit for good measure; the following September, Elon Musk revealed SpaceX's plan to send a Japanese billionaire around the moon on an even bigger rocket, in 2023—a dangerous mission that Musk warned was "not a sure thing"; and preparations were well underway for SpaceX to ferry two NASA astronauts to the ISS, which they did successfully in May 2020, becoming the first private space outfit to send humans into orbit and to safely return them to Earth.

Virgin clung to its niche: it had a spaceship flown by *pilots*, perhaps the best in the world. Divergent thinkers. Calm under pressure. Battle-tested. But they were still humans—prone to errors and miscalculations and oversights and distractions, to "skipping a chapter in the hymnal," to deteriorating vision, to bad days. In short, fallible.

And while Branson worked his magic on the floor of the New York Stock Exchange that morning, his lead test pilot, the one who had first flown them into space, and who saved the ship on several occasions, was home, convalescing in bed.

TWO WEEKS LATER, the movers arrived, bubble-wrapping stemware and packing away heirlooms in brown boxes.

Sometimes the packers would holler to Stucky and Agin to ask about some item or other. They were extra careful in Stucky's office with all of those pictures and model airplanes, and even more so when they learned that Stucky flew rocket ships for a living.

What about the droopy birthday balloons, they asked.

Stucky hobbled over; he wore an air-cast on one foot and a Pyrex stump below the knee on the other. Those were left over from his birthday, he said. Garbage.

What about this one, asked the packer, pointing to the deflated red balloon from Alsbury's memorial service.

That one's coming, said Stucky.

He had meant to pack it away earlier. He had taken many of the pictures down himself. He found it therapeutic, an exercise in cataloging memories. Some of the moments captured in those pictures were public and known. But plenty of others remained cloaked in mystery, their backstories locked away in secret files, and some of those moments, it turned out, were ones he had shared with Alsbury on the range.

Dillon drove up from L.A. that week to help with the move. He was going to miss that drive. It had become a kind of cleansing and restorative ritual—leaving the glitz of L.A. to go hang out with his dad in the desert. Dillon joked with his siblings about taking advantage of Stucky in his currently sedated state: "If he's on good painkillers I actually have some Area 51 questions I'd like to ask him."

Every hour, as his house looked less like his home and more like a loading dock, Stucky felt a sort of creeping melancholy. He had no idea what the coming weeks and months and years would entail, whether Virgin could create a lasting business or not.

But the truth of the past pressed in on him.

He remembered moving in, back when he was newly remarried and chasing his astronaut dream, but estranged from his children and never more alone. He had proven himself and remade his life as a husband, as an astronaut, and, most importantly, as a father. He was going to miss this house because he was going to miss the life he made in it. He was already feeling the ache of separation from Sascha, Lauren, and Dillon.

"I love flying with him but I just absolutely love hanging out with him," Stucky said about Dillon. "And that's coming to an end." He knew it was bound to happen anyway, that someday soon Dillon would reveal that he was getting married or that his longtime girlfriend was pregnant. Said Stucky, "His life would change and then it's that cat-and-the-cradle thing, where you just don't hang out that much anymore anyway." Or at least that's what Stucky told himself.

He led Dillon into the garage where his paraglider was still rolled up in a ball from the day he fell. He asked Dillon to help him pack it up properly and they drove to the hillside where they flew that day in August with their new paragliders, because to pack it up they needed to open it up and either kite it or fly it.

Dillon hadn't flown since that day in August. He had really spooked himself. The sight of Stucky's broken feet didn't ease his anxiety. If Dillon didn't get back on the saddle now, he might never get on.

Take it slow, said Stucky.

Dillon put his foot through the harness one leg at a time and tightened the straps and tested his walkie-talkie. All set.

He bundled the paraglider under his arm, carried it to the launch point, and spread it out on the ground.

Stucky watched Dillon start to run, those graceful high knees that reminded Stucky of Dillon's track-and-field days.

And now Stucky was a spectator once again, admiring from afar, as his son took flight with that first step—the one that encounters nothing solid.

AUTHOR'S NOTE

Test Gods is a personal story about men and women at work, so it seems apt that I should take a moment to explain how I work.

Everything in these pages happened; everything between quotation marks was said. I relied on a multitude of sources, including interviews, email records, government files obtained through Freedom of Information Act requests, company records, postaccident investigation reports, cockpit videos, previously published accounts, contemporaneous notes, and many others.

I, moreover, witnessed many of the events and conversations myself. I was effectively embedded with Virgin Galactic for almost four years, an arrangement that dated back to late 2014, shortly after the crash, when I went to my editor at the *New Yorker* and proposed writing a story about the company. He agreed, so long as we could get "real access."

Two weeks later, I flew to Mojave to meet C.J. Sturckow, an old family friend whom I'd not seen in more than twenty years. Sturckow introduced me to Mike Moses. Moses welcomed me in.

The company placed few restrictions on me. I made, in total, sixteen trips to Mojave and conducted hundreds of taped interviews. I sat in on—and recorded, with their permission—meetings and phone calls. I shadowed engineers, followed Stucky and Moses and others around, and accompanied the pilots on training flights. I was around so much that a new hire once confused me for an employee.

After the *New Yorker* piece appeared, in August 2018, my embedded status was revoked. Moses instructed employees that I was no longer embedded and to cease speaking with me. "It will be an active effort to pull back the access, but that's what we need to do," he wrote. Some relationships endured, however, and I continued to gather information. I was invited, with other reporters, to the first two space flights. And in March 2019, Richard Branson invited me to come spend a long weekend with him in the British Virgin Islands; Stephen Attenborough revoked the invitation forty-eight hours before I was scheduled to depart for Tortola.

The notes on the following pages provide a guide for curious readers who want additional information from public sources. They do not, however, provide a guide for curious readers who want to identify confidential sources; a journalist's first priority is to protect his or her sources. That said, these confidential sources—in some cases, individuals; in others, documents and related materials—are not secret: when I finished writing *Test Gods*, I then provided an annotated manuscript, with citations for every factual assertion in the text, to a professional fact-checker who spent months re-reporting the book—re-interviewing sources, consulting documents, re-reading archival material—to ensure its accuracy.

His efforts contributed immensely to the integrity and robustness of this book. Any praise should be shared; any mistakes are mine and mine alone.

NOTES

PROLOGUE: BLUE ZEBRA

2 **"The only way of discovering the limits of the possible"**: Arthur C. Clarke, *Profiles of the Future: An Inquiry into the Limits of the Possible* (New York: Harper & Row, 1962).

2 **"fighting to try to recover"**: Noah Adams interview with Chuck Yeager, NPR, 1985.

3 **"quintessential statement of our fundamental insanity"**: Norman Mailer, *Of a Fire on the Moon* (Boston: Little, Brown, 1970).

3 **"the first of the three entrepreneurs"**: Christine Romans interview with Richard Branson, Goldman Sachs 10,000 Small Businesses Summit in Washington, DC, February 13, 2018.

6 **"an unpleasant thing to think about"**: Mailer, *Of a Fire on the Moon.*

1: BORN TO FLY

9 **John Glenn slipped on a bathrobe**: John Glenn and Nick Taylor, *John Glenn: A Memoir* (New York: Bantam Books, 1999).

9 **His blood pressure measured 120 over 80**: "John Glenn: One Machine That Worked Without Flaw," *Newsweek*, March 5, 1962.

9 **"there wasn't any closer brotherhood"**: Deborah Lutterbeck, "Mercury Astronaut John Glenn Recalls First Orbit Flight, 50 Years Ago," Reuters, February 19, 2012.

9 **subjected to bowel probes . . . Shortly before ten a.m., the countdown reached zero**: Tom Wolfe, *The Right Stuff* (New York: Farrar, Straus and Giroux, 1979).

9 **"Don't be scared"**: Taylor, *John Glenn.*

10 **"What I do is pretty much my business"**: Deke Slaton, Mercury Astronaut Team press conference, Washington DC, April 9, 1959.

10 **"I got on this project"**: Glenn, Mercury Astronaut Team press conference.

10 **"We're underway"**: Wolfe, *The Right Stuff.*

10 **"He's in the hands of the Lord"**: Gay Talese, "50,000 on Beach Strangely Calm as Rocket Streaks Out of Sight," *New York Times*, February 21, 1962.

10 **"Go, baby!"**: Walter Cronkite, CBS, February 20, 1962.

11 **"thousands of small, luminous particles"**: John Glenn, Mercury 6 transcript, February 20, 1962.

11 **he got dinner at the White House:** Joseph A. Esposito, *Dinner in Camelot: The Night America's Greatest Scientists, Writers, and Scholars Partied at the Kennedy White House* (Lebanon, NH: ForeEdge, 2018).

12 **Congregants of his church had come . . . ostracized:** Jeffrey W. Koller, *The Eden Peace Witness: A Collection of Personal Accounts* (Wichita: Jebeko Publishing, 2004).

12 **"a denial of the Christian faith":** Koller, *The Eden Peace Witness.*

12 **"There was never any doubt":** Koller, *The Eden Peace Witness.*

12 **One day, he found an article:** Joseph A. Walker and Dean Conger, "I Fly the X-15, Half Plane, Half Missile," *National Geographic*, September 1962.

13 **"West Point schmuck":** Karen Lambert, "An Angry, Profane Abbie Hoffman Urges Youth to Join His Revolution," *Salina Journal*, April 10, 1970.

13 **threw eggs at the stage:** Lambert, "An Angry, Profane Abbie Hoffman Urges Youth to Join His Revolution."

13 **Two years later Mark picked up:** Kenneth F. Weaver, "To the Mountains of the Moon," *National Geographic*, February 1972.

16 **BORN TO FLY:** Tom Carlin, "Born to Fly," *K-Stater Magazine*, March 1980.

2: FORGER

17 **"I just kept reminding myself":** Alex Beam, "Michael Collins, the Lucky, Grumpy Astronaut," *Boston Globe*, October 22, 2018.

18 **falsely accused of cheating on his entrance exam:** Sam Roberts, "Frank E. Petersen, First Black General in Marines, Dies at 83," *New York Times*, August 26, 2015.

18 **"not a craft but a fraternity":** Tom Wolfe, "The Truest Sport: Jousting with Sam and Charlie," *Esquire*, October 1, 1975.

19 **"officer of the highest caliber":** Mark Stucky personnel files, comments from R. C. Branch, January 1988.

19 **"unlimited career potential":** Mark Stucky personnel files, comments from J. J. Barta, August 1984.

19 **"superior aeronautical abilities":** Mark Stucky personnel files, comments from J. A. Mitchell, May 1989.

19 **"Unquestionably one of the most":** Mark Stucky personnel files, comments from Robert R. Zimmerman, December 1985.

19 **"It came to that, time after time, who could see the farthest":** James Salter, *The Hunters* (New York: Harper & Brothers, 1956).

20 **"Feel that mother go":** Michael Smith, transcript from the Challenger tape, January 28, 1986.

20 **"We mourn seven heroes":** Ronald Reagan, Space Shuttle Challenger Speech from the Oval Office, January 28, 1986.

20 **"We regretted having to inform you":** Letter to Mark Stucky from Duane L. Ross, manager, NASA Astronaut Selection Office, January 19, 1990.

21 **tenth out of seventeen:** Letter to Mark Stucky from Duane L. Ross, manager, NASA Astronaut Selection Office, May 27, 1992.

21 **"Competition for the program was again extremely keen":** Letter to Mark Stucky from Duane L. Ross, manager, NASA Astronaut Selection Office, April 3, 1992.

21 **"the most significant voyage of my life":** Buzz Aldrin with Wayne Warga, *Return to Earth* (New York: Random House, 1973).

22 **"test pilot dreams":** Mark Stucky, "Flying the Blackbird," *Flyer Magazine*, Summer 2010.

22 **None of the other Dryden pilots:** Tom Tucker, "The Eclipse Project," NASA History Division, 2000.

22 **"expertise, diligence, and impressive professionalism . . . greatly increased":** NASA Nonsupervisory Appraisal, September 9, 1997.

22 **"incredible curves . . . tall wide stance . . . her skin sinisterly-covered in black":** Stucky, "Flying the Blackbird."

23 **"I like to do far-out things with airplanes"**: Miles O'Brien interview with Burt Rutan, live segment on private space exploration, CNN, May 9, 2003.

23 **He won so many remote-control airplane competitions**: Vera Foster Rollo, *Burt Rutan: Reinventing the Airplane* (Baltimore: Maryland Historical Press, 1991).

23 **"twice as tall"**: Irene Mona Klotz, "Rutan Meets His Rocket Heroes," BBC.com, October 25, 2004.

24 **"a blot upon fair Nature's face"**: Edmund J. Carpenter, "Mojave, the Phantom City," *Godey's Magazine* 131 (July 1895).

24 **Lethal pit vipers slithered among the creosote bushes**: Julian Guthrie, *How to Make a Spaceship: A Band of Renegades, an Epic Race, and the Birth of Private Spaceflight* (New York: Penguin Press, 2016).

24 **"driving out the tramps like rabbits"**: "Man Hunt in Hobo Haunts on Desert," *Los Angeles Times*, March 16, 1908.

24 **dry his grapes and sell them as raisins**: Rollo, *Burt Rutan*.

24 **"a creative battering ram"**: Andy Meisler, "Slipping the Bonds of Earth and Sky," *New York Times*, August 3, 1995.

24 **a catamaran that won the 1988 America's Cup**: Daniel Alef, *Burt Rutan: Aeronautical and Space Legend* (Jersey City, NJ: Titans of Fortune Publishing, 2016).

24 **a white hexagonal pyramid**: Jim Schefter, "21st Century Pyramid," *Popular Science*, November 1989.

25 **"It is disappointing . . . make the possible impossible"**: Mark Stucky, NASA Request for Personnel Action, Part E: Employee Resignation/Retirement, September 3, 1999.

3: A BEAUTIFUL BALL

28 **"no safety of flight issue"**: Rowland White, *Into the Black: The Extraordinary Untold Story of the First Flight of the Space Shuttle Columbia and the Astronauts Who Flew Her* (New York: Atria Books, 2017).

28 **open gash and details from the Columbia descent**: Columbia Accident Investigation Board Report, vol. 1, August 2003.

29 **"strong tactical background . . . multiple weapons schools"**: Mark Stucky, "Application for Extended Active Duty with the United States Air Force," February 23, 2003.

29 **krypton and xenon**: Interview with Jonathan McDowell.

30 **"Don't start talking shit"**: Email provided to author.

30 **"Don't know if you've seen CNN"**: Email provided to author.

30 **"use a leash"**: Mark Stucky, "Forger News" email newsletter, August 10, 2004.

30 **a harrowing low-altitude helicopter ride**: Mark Stucky, "Forger News" email newsletter.

4: GOD IS HERE

32 **"more like a Midwestern grain silo"**: Frank Sweeney and Glennda Chui, "Space Flight on the Cheap: Roton to Launch New Industry in Commercial Rockets," *San Jose Mercury News*, March 2, 1999.

32 **browbeating the NASA head**: Guthrie, *How to Make a Spaceship*.

33 **"Unless guys like me go out and do this"**: *Black Sky: The Race for Space*, directed by Gail Willumsen, Sandy Guthrie, Jill Shinefield, and Scott B, Vulcan Productions, 2004.

33 *God is here*: Guthrie, *How to Make a Spaceship*.

33 **"Your worst outcome is an error message"**: Paul Allen, *Idea Man: A Memoir by the Cofounder of Microsoft* (New York: Portfolio, 2011).

33 **"confidence in nonsense":** *Black Sky.*

34 **"I do not claim to have solved any of the mysteries":** John B. Alexander, *UFOs: Myths, Conspiracies, and Realities* (New York: St. Martin's Press, 2011).

34 **without a wind tunnel:** Alef, *Burt Rutan.*

34 **throwing foam models:** Guthrie, *How to Make a Spaceship.*

35 **"Been to two goat ropings":** Alan Boyle, "First Private Space Pilot 'Ready to Go,'" MSNBC .com, June 21, 2004.

35 **"He flat didn't fly the airplane":** Eric Adams, "The New Right Stuff," *Popular Science*, November 1, 2004.

36 **"Wow, it's quiet up here":** Guthrie, *How to Make a Spaceship.*

37 **"The sky of Mojave is very big":** Guthrie, *How to Make a Spaceship.*

37 **Binnie went on *The Late Show*:** *The Late Show with David Letterman*, October 7, 2004.

38 **"If the desert is turning red":** "Rocket Man," Forbes.com, July 29, 2006.

5: FOGGLES

39 **a Beechcraft flew over and details from Area 51 founding:** Gregory W. Pedlow and Donald E. Welzenbach, *The CIA and the U-2 Program, 1954–1974*, CIA, Center for the Study of Intelligence, 1998.

39 **tedious wasteland:** Trevor Paglen, *Blank Spots on the Map: The Dark Geography of the Pentagon's Secret World* (New York: Barkley, 2010).

39 **"soul-shattering silence":** Freeman Dyson, *Disturbing the Universe* (New York: Harper & Row, 1979).

39 **tiny dorms:** William B. Scott, "The Truth Is Out There," *Air & Space Magazine*, September 2010.

39 **The fleet of unmarked jets, named "JANET":** Trevon Milliard, "Nevada's Best-Kept Secrets Endure at Area 51," *Las Vegas Review-Journal*, August 17, 2014.

40 **a half dozen turboprops and details from the program:** Stephanie M. Smith, "Deployed Flight Test of the Iraqi Air Force Comp Air 7SLX (CA-7)," Air Force Test Center History Office, February 2014.

40 **"incredible resourcefulness and real courage":** Citation to Accompany the Award of the Lieutenant General Bobby Bond Memorial Aviator Award for 2005 to Lieutenant Colonel Mark P. Stucky.

41 **Everything about the planes came under strict secrecy:** Ben Rich and Leo Janos, *Skunk Works: A Personal Memoir of My Years at Lockheed* (Boston: Little, Brown, 1994).

41 **satellite overflights:** Scott, "The Truth Is Out There."

41 **Soviet intelligence analysts:** Rich and Janos, *Skunk Works.*

41 **"It was my responsibility to be there for my family":** Email provided to author.

6: THE REBEL BILLIONAIRE

43 **"We're not into music":** John Robb, *Punk Rock: An Oral History* (London: Ebury Press, 2006).

44 **Vicious would later admit to killing:** (Vicious subsequently recanted.) Alan Parker, *Sid Vicious: No One Is Innocent* (London: Orion, 2008).

44 **"You are either going to be a millionaire or are going to jail":** Michael Specter, "Branson's Luck," *New Yorker*, May 14, 2007.

44 **"I won't let silly rules":** Richard Branson, *Screw It, Let's Do It: Lessons in Life* (London: Virgin Books, 2006).

44 **"attack a marketplace":** Jason Ankeney, "Richard Branson on Building an Empire," *Entrepreneur Magazine*, June 2012.

45 **"self-made and self-deprecating man who"**: Tom Bower, *Branson: Behind the Mask* (London: Faber & Faber, 2014).

45 **lists of both British heroes and scoundrels**: Georgina Cooper, "Richard Branson Beats Jesus in British Role Model Poll," Reuters, February 28, 2008.

46 **"We will build another boat and try again"**: Jo Thomas, "Powerboat Sinks Just Hours from Breaking Atlantic Speed Record," *New York Times*, August 16, 1985.

46 **"They're building a fucking spaceship"**: Richard Branson, *Finding My Virginity* (New York: Portfolio/Penguin, 2017).

46 **"I hope in five years a reusable rocket"**: "Richard Branson's Latest Idea Flies High," *Cedar Rapids Gazette*, May 24, 1999.

47 **"Silicon Valley for the new industry"**: John Johnson Jr., "Mojave: Edge of the Final Frontier," *Los Angeles Times*, October 2, 2007.

48 **a Japanese TV station paid**: Kathy Sawyer, "Japanese to Become First Journalist in Space," *Washington Post*, November 12, 1990.

48 **"moments of terror . . . pure joy . . . old Swiss clockmaker"**: Adam Fisher, "Very Stunning, Very Space, and Very Cool," *MIT Technology Review* 112, no. 1 (December 22, 2008).

48 **they billed the Japanese TV station**: Thomas Ginsburg, "Japanese Company Critical of Soviet Business Practices," Associated Press, December 8, 1990.

48 **Tito inherited his ticket from**: Fisher, "Very Stunning, Very Space, and Very Cool."

48 **NASA officials barred him**: Peter Jennings, "NASA and Russia Feud over Millionaire Dennis Tito Going Along for Ride to International Space Station," *World News Tonight* transcript, March 20, 2001.

48 **later threatened to bill him . . . listening to opera on his headphones**: Todd Venezia, "NASA to Space Tourist: You Owe Us," *New York Post*, May 6, 2001.

48 **"I just came back from paradise"**: Marcus Warren, "Trip Was Out of This World, Says First Space Tourist," *Telegraph* (UK), May 7, 2001.

48 **"the overview effect"**: Frank White, *The Overview Effect: Space Exploration and Human Evolution* (Boston: Houghton Mifflin, 1987).

48 **"feeling of unity"**: David Bryce Yaden et al., "The Overview Effect: Awe and Self-Transcendent Experience in Space Flight," *Psychology of Consciousness: Theory, Research, and Practice* 3, no. 1 (March 2016): 1–11.

48 **"very, very emotional"**: Terry Gross interview with Alan Shepard, NPR, July 15, 1994.

49 **"That's here. That's home"**: Carl Sagan, *Pale Blue Dot* (New York: Random House, 1994).

49 **"Within five years"**: "British Tycoon Branson Unveils Plan for Commercial Space Flights," Agence France-Presse, September 27, 2004.

50 **"poison the process"**: Burt Rutan speech at Stanford University Engineering and Aeronautics Department, February 2003.

50 **"real stimulating projects"**: Rollo, *Burt Rutan*.

7: HIDDEN TIGERS

52 **Colby's high school mentor**: The mentor was Elliot Ring, a late rocket scientist who was the head of the Titan I and Titan III missile development programs and worked on the United States' first hybrid motor program.

52 **"While I am thrilled to be the one and only 'Rocket Guy'"**: Luke Colby journal entry.

53 **"I am not just a thrill seeker"**: Glen May, "The Barnstorming Air Show Rocket," realrocketman .tripod.com.

53 **"crazy, hopeless romantic"**: Luke Colby journal entry.

53 **"wide open highway we must travel"**: "The Barnstorming Air Show Rocket."

53 **"a group of excellent people . . . bring us back from the gutter":** Luke Colby journal entry.

54 **the SCUM truck:** State of California, Department of Industrial Relations, Division of Occupational Safety and Health, Investigation No. 201089349, Narrative Summary, February 22, 2008.

55 **blistering hot:** State of California, Department of Industrial Relations, Division of Occupational Safety and Health, Field Documentation Worksheet, Inspection No. 310821103, July 27, 2007.

55 **about a dozen of them standing near the fence:** State of California, Department of Industrial Relations, Division of Occupational Safety and Health, Investigation No. 201089349, Narrative Summary, February 22, 2008.

56 **"raised serious questions":** Jeff Hecht, "Spaceship Company Fined for Safety Violations," NewScientist.com, January 22, 2008.

56 **"he could be on his way out":** Luke Colby journal entry.

57 **"hidden tigers":** Luke Colby journal entry.

8: EVIL AIR

59 **"The venue out here isn't as large":** Email provided to author.

59 **"elegant simplicity of the sport . . . flight of the birds":** Mike Meier and Mark P. Stucky, *Paragliding: A Pilot's Training Manual* (Orange, CA: Wills Wing, 2004).

60 **"You asked me about what I have to offer":** Email provided to author.

61 **"natural enemies of pilots":** Glenn and Taylor, *John Glenn: A Memoir*.

62 **"My wife had no appreciation":** Mark Stucky, "Hang in There—Epilogue," *Hang Gliding and Paragliding Magazine*, January 2010.

64 **"discontinue communication":** Text message provided to author.

64 **"stake in the heart":** Email provided to author.

64 **"The only way I can see that I would be so shutout":** Facebook message provided to author.

9: THE PARTY

66 **"bread and butter":** David Snow interview with Mark Patterson, July 9, 2012.

66 **A flood of applications:** Adam Higginbotham, "Up: The Story Behind Richard Branson's Goal to Make Virgin a Galactic Success," *Wired*, March 2013.

66 **"selling a dream":** Kenny Kemp, *Destination Space: Making Science Fiction a Reality* (London: Virgin Books, 2007).

66 **"I had to warn clients that products could fall as well as rise":** Robert Watts, "Fore Street," *Sunday Telegraph* (UK), January 16, 2005.

67 **"open for business by the beginning of 2005":** Virgin Galactic, "Virgin Group Sign Deal with Paul G. Allen's Mojave Aerospace," press release, September 27, 2004.

67 **"The flow of sophisticated weapons":** Clifford P. Case, Congressional Record—Senate, February 4, 1976.

67 **"If you are going to be courageous":** "Game Changers Gather for 5th Edition of MIPIM UK at Critical Time for UK Property," MIPIM UK press release, October 19, 2018.

68 **"Cosmic Brand and Marketing":** "DediPower Helps Send Virgin Galactic Supersonic," press release, December 17, 2009.

68 **Necker was a tropical paradise:** Specter, "Branson's Luck."

68 **"We will dub the event '2008'":** Internal correspondence provided to author.

69 **"Business as usual":** Alicia Chang, "Publicity-Seeking Virgin Galactic Keeps Low Profile After Mojave Desert Explosion," Associated Press, August 26, 2007.

69 **"a Chevrolet's functional reliability":** Rich and Janos, *Skunk Works*.

69 **"One bolt falls off and you die"**: Higginbotham, "Up: The Story Behind Richard Branson's Goal to Make Virgin a Galactic Success."

70 **Jornada del Muerto**: "The Dead Man of the Jornada: El Hombre Muerto de La Jornada," El Camino Real de Tierro Adentro National Historic Trail, Bureau of Land Management, nps.gov.

70 **"New Mexico will be known around the world . . . a chance to go home"**: Leonard David, "Virgin Galactic Partners with New Mexico on Spaceport," Space.com, December 14, 2005.

70 **"definitely one of the coolest"**: "Governor's Remarks at Virgin Galactic's Unveiling of Space-ShipTwo," States News Service, December 7, 2009.

71 **"Isn't that the sexiest spaceship ever?"**: John Johnson Jr., "A Giant Step for Space Tourism," *Los Angeles Times*, December 8, 2009.

71 **ice bar, taking shots of vodka**: Tina Forde, "Safety Training Averts Disaster at SpaceShipTwo Rollout," *Tehachapi News*, December 11, 2009.

72 **"There is no way we are going"**: Branson, *Finding My Virginity*.

72 **ordering an evacuation**: Forde, "Safety Training Averts Disaster at SpaceShipTwo Rollout."

72 **"kind of ironic"**: Email provided to author.

10: ANGEL'S WINGS

74 **"so smooth that it looks like someone drew it with a pen"**: Luke Colby journal entry.

74 **"government-funded boondoggles"**: "Statement from Andrew Beal," Beal Aerospace Technologies, Inc., October 23, 2000.

74 **"high-paid assassin"**: Ashlee Vance, *Elon Musk: Tesla, SpaceX, and the Quest for a Fantastic Future* (New York: HarperCollins, 2015).

74 **nearly burned down an island . . . "took their lumps"**: Vance, *Elon Musk*.

75 **dusty Soviet rocket manual**: Vance, *Elon Musk*.

75 **"building a Ferrari for every launch"**: Vance, *Elon Musk*.

76 **"Arrogance got more pilots in trouble"**: Chuck Yeager and Leo Janos, *Yeager* (New York: Bantam Books, 1985).

76 **"Life and death is often quite random in nature"**: Mark Stucky, "Forger News" email newsletter.

77 **"It *can* happen to you"**: Mark Stucky (guest), "Cloudbase Mayhem" podcast, episode 82, December 26, 2018.

78 **"The balloon is inflated with hydrogen"**: "The Santos-Dumont Balloon," *Scientific American*, August 10, 1901.

78 **"the engine conditions which will give the maximum economy"**: Royal Aeronautical Society, *The Aeronautical Journal*, vol. 48 (1944).

78 **"The 'envelope' was a flight-test term"**: Wolfe, *The Right Stuff*.

80 **"angel's wings"**: *Black Sky*.

11: THE GAMMA TURN

82 **"We dissect what it is that is going to scare us"**: Terry Gross interview with Chris Hadfield, NPR, October 30, 2013.

82 **"You can see the future from Mojave"**: Stuart O. Witt, Hearing on Commercial Space, Subcommittee on Space, Committee on Science, Space, and Technology, US House of Representatives, November 20, 2013.

83 **"We need to make some bold moves . . . Fire them as soon as possible"**: Email provided to author.

83 **"Delay is a strange word"**: "Virgin Galactic Space Tourism Could Begin in 2013," BBC News, October 26, 2011.

83 **"Things are going incredibly well"**: Robin McKie, "Destination Outer Space," *Observer Magazine*, June 17, 2012.

83 **"exciting brand . . . build momentum"**: Lucy Tesseras, "Space Age Thinking," *Marketing Week*, December 11, 2013.

83 **"democratization of space"**: Charisse Jones, "As Shuttle Fades, the Race for Space Tourism Is Flying High," *USA Today*, July 25, 2011.

83 **"being on the inside of a blast furnace"**: Fisher, "Very Stunning, Very Space, and Very Cool."

84 **"predicated on false or deliberately"**: Doug Messier, "London Sunday Times Story About Cracks in WhiteKnightTwo's Wings," ParabolicArc.com, May 11, 2014.

84 **"raise his voice or lose his cool . . . data computer . . . turn any sky into a series of numbers"**: "Co-pilot Who Died in Virgin Galactic Crash Hailed as 'Renaissance Man,'" Associated Press, November 2, 2014.

86 **"I've taken flights, rented cars"**: Facebook message provided to author.

87 **"Once you experience that"**: Dave Davies interview with David Carr, NPR, August 12, 2008.

88 **"icebergs of mystery"**: Mailer, *Of a Fire on the Moon*.

88 **"driving on bad shock absorbers"**: Yeager and Janos, *Yeager*.

89 **"not down on any map"**: Herman Melville, *Moby-Dick* (New York: Harper & Brothers, 1851).

92 **while the second man in space**: Wolfe, *The Right Stuff*.

92 **The award cited Stucky's**: Remarks at Iven C. Kincheloe award ceremony, September 28, 2013.

93 **"I've got a cool flight"**: Facebook messages provided to author.

12: THEY'RE GONE

95 **"I always love Christmas presents"**: Peter Diamandis interview with Richard Branson, October 4, 2014.

95 **on his way to train in a centrifuge**: Branson, *Finding My Virginity*.

96 **"any rational person would be slightly nervous"**: Alan Boyle, "First Family of Space: Richard Branson and Kids Will Blaze New Trail," NBCNews.com, November 8, 2013.

99 **went through the bailout procedure . . . feeling for the d-ring . . . and the buckles**: The accident was documented extensively by the National Transportation Safety Board's investigation, which published hundreds of pages of reports, interviews, and findings. National Transportation Safety Board, Office of Aviation Safety factual report, interview Peter Siebold, "Survival Factors: Attachment 3—Interview Summaries," January 16, 2015.

13: WRECKAGE

101 **"paper fluttering in the wind"**: "Survival Factors: Attachment 3—Interview Summaries," Peter Siebold, National Transportation Safety Board, Office of Aviation Safety, January 16, 2015.

102 **A manager picked up the phone**: National Transportation Safety Board, Office of Aviation Safety factual report, interview with Ben Diachun, "Survival Factors: Attachment 3—Interview Summaries," February 19, 2015.

102 **Stucky pulled the Blue Zebra binder**: National Transportation Safety Board, Office of Aviation Safety factual report, interview with Mark Stucky, "Survival Factors: Attachment 1—Interview Summaries," November 3, 2014.

102 **pages of procedures**: National Transportation Safety Board, Office of Aviation Safety factual report, interview with Charlie Clark, "Survival Factors: Attachment 3—Interview Summaries," February 25, 2015.

103 **"scene out of a car bomb"**: Doug Messier, "SpaceShipTwo Crash Eyewitness," YouTube, June 8, 2015.

104 **"hunting" for cops:** Steven Mayer, "Alleged Bomb Builder Shakes Up Kern County Desert Community," Bakersfield.com, September 13, 2016.

104 **"Because it was a test flight":** Jeff Foust, "NTSB Begins SpaceShipTwo Investigation as Pilot Identified," Spacenews.com, November 1, 2014.

105 **"courage, intelligence, and good nature . . . even more resolve and commitment":** Branson, *Finding My Virginity.*

105 **"Bad day . . . On my way":** Branson, *Finding My Virginity.*

105 **"the world's most expensive roller coaster . . . boondoggle thrill ride":** Adam Rogers, "Space Tourism Isn't Worth Dying For," *Wired*, October 31, 2014.

106 **"The rocket stopped . . . They took this pilot's life":** Joël Glenn Brenner interview on CNN Newsroom, October 31, 2014.

106 **"We fell short":** Kaytlyn Leslie, "Both Test Pilots in Virgin Galactic Crash Graduated from Cal Poly," *San Luis Obispo Tribune*, November 1, 2014.

106 **"can't afford to lose anybody":** Leigh Buchanan, "Big Ideas Also Mean Big Risks," inc.com, November 6, 2012.

107 **"Fire in the cockpit":** "Report of Apollo 204 Review Board," National Aeronautics and Space Administration, April 5, 1967.

107 **"the way things go in our business":** Terry Gross interview with Alan Shepard, NPR, July 15, 1994.

107 **SPACE TOURISM ISN'T WORTH DYING FOR:** Rogers, "Space Tourism Isn't Worth Dying For."

107 **"hype factory":** Doug Messier, "Mojave Journal: The Ansari X Prize's Awkward Family Reunion," ParabolicArc.com, October 5, 2015.

107 **"a great friend and a great spaceship":** Marky Stucky comment, "Whitesides Vows to Stay the Course, Defends Virgin Galactic's Approach to Safety," by Doug Messier, ParabolicArc.com, November 15, 2014.

108 **"We've received countless messages . . . no mission more remarkable":** Mark Stucky, Memorial Service for Mike Alsbury, November 13, 2014.

109 **"Glad you are OK":** Facebook message provided to author.

109 **"In the absence of hearing":** Facebook message provided to author.

109 **"Son ignores father and refuses":** Facebook message provided to author.

14: THE RED BALLOON

110 **"We thought we had it licked":** Marcia Dunn, "Shuttle Launch Off Until End of Month to Fix Leak," Associated Press, November 5, 2010.

110 **"I can't see how anyone who comes down":** Donna Leinwand Leger, "'A Fitting End' to a Stellar Ride; NASA, Nation Share Pride at Last Liftoff," *USA Today*, July 25, 2011.

111 **"The work force down here":** Todd Halvorson, "KSC Workers Focus Despite Uncertainty," *Florida Today*, January 31, 2010.

111 **"I'm going to take another shot":** Irene Klotz, "NASA's Space Shuttle Operations Chief Heading to Virgin," Reuters, October 11, 2011.

113 **It, like other early telescopes:** Neil deGrasse Tyson and Avis Lang, *Accessory to War: The Unspoken Alliance Between Astrophysics and the Military* (New York: W. W. Norton, 2018).

113 **"Give me a warm day and I'll fly":** Guthrie, *How to Make a Spaceship.*

15: A RESEMBLANCE

118 **"Keep that man away":** Mailer, *Of a Fire on the Moon.*

118 **A mountain lion attacked:** Leo Stallworth, "Mountain Lion Caught After Pouncing on Palmdale Mechanic," *ABC 7 Eyewitness News*, August 28, 2015.

119 **"estranged backcountry stepchild"**: Jennifer Swann, "The Last Artists' Haven in Los Angeles," *Curbed Los Angeles*, December 15, 2016.

119 **a list of the fifty most miserable:** James Pasley and Angela Wang, "The 50 Most Miserable Cities in America, Based on Census Data," *Business Insider*, September 29, 2019.

119 **Child abuse hotlines struggled:** Garrett Therolf, "Why Did No One Save Gabriel?" *Atlantic*, October 3, 2018.

119 **"No one can get out of here":** Jaime Lowe, "The Incarcerated Women Who Fight California's Wildfires," *New York Times Magazine*, August 31, 2017.

119 **"Not in a thousand years would":** David McCullough, *The Wright Brothers* (New York: Simon & Schuster, 2015).

120 **"I wouldn't have gone ahead":** Branson, *Finding My Virginity*.

16: JOUSTING

122 **The article was by Tom Wolfe:** Wolfe, "The Truest Sport: Jousting with Sam and Charlie."

125 **"shrouded coffins":** William Claiborne, "Saddam Calls Bush a 'Liar,' Rejects Pullout; Iranian Leader Repeats Demand for Iraq to Relinquish Kuwait," *Washington Post*, August 17, 1990.

125 **"superb airmanship and undaunted":** Distinguished Flying Cross citation, from Sean O'Keefe, acting secretary of the Navy, December 11, 1992.

125 **"You'd rather be here":** Robert E. Schmidle letter to Sonny Carter, February 13, 1991.

126 **the plane rolled hard and suddenly:** "Aircraft Accident Report: Atlantic Southeast Airlines, Inc., Flight 2311," National Transportation Safety Board, April 28, 1992.

17: WHALE LINES

128 **"magical, sometimes horrible . . . whale lines":** Melville, *Moby-Dick*.

130 **"the military guy":** Aaron Lammer, Evan Ratliff, and Max Linsky, "#46: Nicholas Schmidle," Longform Podcast, June 19, 2013.

131 **shelling Muslims:** Chuck Sudetic, "Serbs Press Attack on Muslims in Gorazde," *New York Times*, April 4, 1994.

131 **The UN commander warned Serb general Ratko Mladić:** Chuck Sudetic, "Conflict in the Balkans: The Overview; U.S. Planes Bomb Serbian Position for a Second Day," *New York Times*, April 12, 1994.

132 **Serbs were raping and massacring and burning babies:** "Bosnian Serb Jailed for 20 Years for Burning Bosniaks to Death," *Al Jazeera*, October 30, 2019.

132 **"I can't see that son of a bitch" and other details of the flight:** Cockpit video, April 4, 1994.

133 **his jet needed to be rearmed and other background details of the mission:** Fred H. Allison, "Thunderbolts Rain Down on the Serbs During Operation Deny Flight," *Fortitudine* 37, no. 3, (2012) 36–37.

133 **"there will be escalation":** Sudetic, "Conflict in the Balkans."

133 **shot out of the sky:** John F. Harris and Jonathan C. Randal, "NATO Plane Shot Down in Battle for Gorazde," *Washington Post*, April 17, 1994.

134 **"I am pretty tired of being estranged":** Facebook messages provided to author.

134 **"You are Marine kids":** Pat Conroy, *The Great Santini* (New York: Random House, 1976).

18: BLANK PAGES

137 **"miracle material":** Alfred Friendly, "Edison's Filament Lightens 'Steel,'" *Washington Post*, January 4, 1970.

138 **"You have to understand how hardware behaves"**: Alex Ruppenthal, "Former NASA Engineer's Interesting Take on Women in Science," WTTW.com, May 11, 2017.

139 **"20 percent of all engineering majors"**: Mark Crawford, "Engineering Still Needs More Women," The American Society of Mechanical Engineers, September 12, 2012.

139 **"I was sometimes the only woman"**: Ruppenthal, "Former NASA Engineer's Interesting Take on Women in Science."

139 **There were almost twenty rocket companies**: Witt, Hearing on Commercial Space, Subcommittee on Space, Committee on Science, Space, and Technology, US House of Representatives.

140 **"Rockets are a fundamentally difficult thing"**: Rebecca Boyle, "Elon Musk Reveals Cause of SpaceX Explosion," PopularScience.com. July 20, 2015.

141 **"trying to finish one last big job"**: Email provided to author.

19: STARMAN

147 **drove home with their son's ashes:** Gabrielle Burkhart, "Grieving Couple Pleads for Return of Son's Stolen Ashes," KRQE, March 1, 2016.

149 **"I've already got what I paid for"**: Elizabeth Culliford, "Cufflinks and the Caribbean: How Virgin Galactic Kept Space Tourists' Interest and Money," Reuters, April 12, 2019.

153 **"one-man publicity circus"**: Geraldine Fabrikant, "Of All That He Sells, He Sells Himself Best," New York Times, June 1, 1997.

153 **"surrounded by the stars"**: Branson, Finding My Virginity.

155 **"We're thoroughly enjoying spending"**: Culliford, "Cufflinks and the Caribbean."

20: JITTERS

160 **aced a national flying contest:** "NIFA Safecon Winners," Flying Magazine, August 1990.

161 **In 1953, when the Air Force:** Teresa Sharp, "In Quest of Answers to X-2 Crash; Author Delving into Lake Tragedy of '53 That Left Two Dead and the Bell Test Plane Lost," Buffalo News, February 4, 2012.

161 **"fine art of chewing ass"**: Todd C. Ericson, "Toward a Fail-Safe Air Force Culture: Creating a Resilient Future While Avoiding Past Mistakes," Air War College, Maxwell Paper No. 66, Air University, October 2012.

163 **"a stricken instant, a cauldron of adrenaline"**: Mailer, Of a Fire on the Moon.

21: THE WALNUT

167 **"The love of flying demands the attention"**: Mailer, Of a Fire on the Moon.

167 **"arcane micrometeorological matters"**: Mark Stucky, Memorial Service for Casey Donohue, November 22, 2012.

168 **Michael Phelps swam and won:** Karen Crouse, "Five Golds for Phelps, Three to Go," New York Times, April 12, 2008.

169 **"What the hell"**: Text message provided to author.

169 **"People make a fair bit of fun"**: Marcia Dunn, "Storms NASA's Only Worry for Shuttle Launch," Associated Press, July 10, 2009.

171 **Before Branson went up he drank a cup:** Branson, Finding My Virginity.

172 **elevated intracranial pressure, swollen:** Justin S. Lawley et al., "Effect of Gravity and Microgravity on Intracranial Pressure," Journal of Physiology 595, no. 6 (March 15, 2017): 2115–27.

172 **chromosomal changes:** Francine E. Garrett-Bakelman et al., "The NASA Twins Study: A Multidimensional Analysis of a Year-Long Human Spaceflight," *Science* 364, no. 6436, DOI:10.1126/science.aau8650, (April 12, 2019).

172 **blood clots in the jugular vein:** Karina Marshall-Goebel et al., "Assessment of Jugular Venous Blood Flow Stasis and Thrombosis During Spaceflight," *JAMA Network Open* 2, no. 11 (2019): e1915011.

175 **"all conscious connection with the past":** Charles Lindbergh, *The Spirit of St. Louis* (New York: Scribner, 1953).

22: GRADATIM FEROCITER

179 **"potentially hazardous outside":** Text messages provided to author.

180 **second-strangest astronaut:** Todd Halvorson, "NASA's Space Shuttle: Cheers to 25 Years from a Veteran Space Reporter," Space.com, April 12, 2006.

182 **Charles Darwin's grandfather used to:** Nicholas J. Wade, "The Original Spin Doctors—The Meeting of Perception and Insanity," *Perception* 34 (2005): 253–60.

185 **"Slow is smooth and smooth":** Cecilia Kang, "Blue Origin, Jeff Bezos's Moon Shot, Gets First Paying Customer," *New York Times*, March 7, 2017.

185 **"You can't cut any corners":** Steve Taylor interview with Jeff Bezos, Pathfinder Awards, The Museum of Flight, October 3, 2016.

185 **$215 million a day:** Julia Limitone, "Amazon Chief Jeff Bezos Rakes in $215 Million a Day," FoxBusiness.com, October 3, 2018.

186 **"plagued with dated paradigms":** Jack Fischer, Presentation at the Space Power Workshop, April 2009.

187 **"muffled their launches to the point of obscurity":** Christian Davenport, "The Unsung Astronauts," *Washington Post*, June 15, 2018.

187 **"a sleeping dragon waking":** Don Pettit, "What's a Soyuz Launch Like?," AirSpaceMag.com, January 11, 2012.

23: ARTIFACTS

188 **"We are hopefully about three months":** Richard Branson, Nordic Business Forum interview, October 2, 2017.

188 **He wore lipstick for one promotional:** "Sir Richard Branson Wears Lipstick and Red Skirt as He Dresses as Flight Attendant Following Bet with Tony Fernandes," *Evening Standard*, May 12, 2013.

188 **"a bizarre sexual allergy":** Specter, "Branson's Luck."

189 **"We have reached the point where":** Email provided to author.

24: FULL DURATION

190 **"this team of innocents":** Email provided to author.

195 **"I don't feel we are being restrictive":** Email provided to author.

195 **"Your shotgun e-mail makes":** Email provided to author.

195 **"To paraphrase Harrison Storms":** Email provided to author.

197 **"I often say that rockets":** Email provided to author.

197 **"It takes a Real Test Pilot":** Email provided to author.

197 **"I've got some cool SS2 fights":** Email provided to author.

199 **"You're bug-eyed, thrilled to your toes":** Yeager and Janos, *Yeager.*

25: SLIPPING THE SURLY BONDS

209 **"time dilation":** Walker and Conger, "I Fly the X-15, Half Plane, Half Missile."

26: MILLION-DOLLAR VIEW

216 FORGER CAN'T GET IT UP: Email provided to author.
218 **"subtleness of the sea":** Melville, *Moby-Dick.*
218 **Three years before Neil Armstrong landed:** Barton C. Hacker and James M. Grimwood, *On the Shoulders of Titans: A History of Project Gemini* (Washington, DC: Scientific and Technical Information Office, National Aeronautics and Space Administration, 1977).
219 **"The guy was brilliant":** "First Man on the Moon," *Nova*, PBS, December 3, 2013.

27: DEFINING SPACE

222 **"If we do not present this year":** Email provided to author.
226 **"The lucky boots":** Jeff Bezos (@JeffBezos), "The Lucky boots worked again," tweet, April 29, 2018.
226 **In 1957, an American named Andrew Haley:** Thomas Gangale, *How High the Sky?* (Boston: Brill Nijhoff, 2018).
227 **McDowell published an article:** Jonathan C. McDowell, "The Edge of Space: Revisiting the Karman Line," *Acta Astronautica* 151 (October 2018): 668–77.
227 **Bell pilots discussed . . . C-17:** 2018 Society of Experimental Test Pilots Symposium program, September 27, 2018.

28: DAGOBAH

230 **"Heads up":** Email provided to author.
231 **"Biggest science fraud in history":** Email provided to author.
233 **"Mars has an atmosphere":** New research shows that the moon does, in fact, have a scant atmosphere. "Is There an Atmosphere on the Moon?" NASA.com, April 12, 2013.
235 **In 2018, they spent $118 million:** Virgin Galactic FY 2018 Financial Results.
235 **"any place on the planet":** David Goldfein, Air Power Conference Keynote Address, July 17, 2017.
235 **"I'm heartened to see Sir Richard":** Email provided to author.

29: FORTUNE COOKIES

239 **"magic is just someone":** Chris Jones, "The Honor System," *Esquire*, September 17, 2012.
240 **"some specifics on the beta":** Email provided to author.
242 **"amiable strangers":** Michael Collins, *Carrying the Fire* (New York: Farrar, Straus and Giroux, 1974).
243 **cubes concocted in labs for maximum:** Mailer, *Of a Fire on the Moon.*
243 **Christmas ornaments, autograph sittings:** Chris Chandler and Andy Rose, "After Walking on Moon, Astronauts Trod Various Paths," CNN, July 17, 2009.

30: THOROUGHBRED

251 **"surrendered words about as happily":** Mailer, *Of a Fire on the Moon.*
253 **"I even forget my lines sometimes":** Natalie Clarkson, "Richard Branson: Learning to Deal with Pressure," Virgin.com, January 6, 2016.

31: THE SPACE MIRROR

261 **"I don't have a death wish"**: "Fueled by Death Cast 141—ASTRONAUT MARK 'FORGER' STUCKY," Death Wish Coffee, August 29, 2019.

262 **"I held the strong belief that there"**: Mark Stucky, "A Hitchhiker's Guide to Future Planning," Salina South High School commencement address, Salina, Kansas, May 19, 2019.

263 **Branson was about to go on CBS**: Interview with Richard Branson and Kevin Plank, *CBS This Morning*, January 24, 2019.

266 **"Science happens not because governments"**: Terry Gross interview with Neil deGrasse Tyson, NPR, September 17, 2018.

270 **"I'm still not ready to accept"**: James Dean, "KSC Memorial May Honor More Than Just Government Astronauts," *Florida Today*, May 29, 2019.

32: WINGS

276 **"If you notice I've gone AWOL"**: Richard Branson, remarks at Smithsonian National Air and Space Museum, February 7, 2019.

280 **Yeager exceeding the specs of the envelope**: J. Terry White, "Final Flight: NF-104A, Ship 3," *White Eagle Aerospace*, December 10, 2019.

284 **"deep, deep black"**: Interview with Beth Moses, Space Stories, Explorers Club, October 9, 2019.

284 **"What am I going to do for the"**: Moses, Space Stories, Explorers Club, October 9, 2019.

EPILOGUE: MAGIC

294 **"He's magic"**: "Squawk on the Street," CNBC, October 28, 2019.

295 **"not a sure thing"**: Christian Davenport, "Elon Musk's SpaceX Plans to Fly a Japanese Billionaire and Several Artists on a Tourist Trip Around the Moon," *Washington Post*, September 17, 2018.

SELECTED BIBLIOGRAPHY

Alef, Daniel. *Burt Rutan: Aeronautical and Space Legend* (Jersey City, NJ: Titans of Fortune Publishing, 2016).

Alexander, John B. *UFOs: Myths, Conspiracies, and Realities* (New York: St. Martin's Press, 2011).

Allen, Paul. *Idea Man: A Memoir by the Cofounder of Microsoft* (New York: Portfolio, 2011).

Bower, Tom. *Branson: Behind the Mask* (London: Faber & Faber, 2014).

Branson, Richard. *Finding My Virginity* (New York: Portfolio/Penguin, 2017).

———. *Screw It, Let's Do It: Lessons in Life* (London: Virgin Books, 2006).

Clarke, Arthur C. *Profiles of the Future: An Inquiry into the Limits of the Possible* (New York: Harper & Row, 1962).

Conroy, Pat. *The Great Santini* (New York: Random House, 1976).

Davenport, Christian. *The Space Barons: Elon Musk, Jeff Bezos, and the Quest to Colonize the Cosmos* (New York: Public Affairs, 2018).

Dyson, Freeman. *Disturbing the Universe* (New York: Harper & Row, 1979).

Esposito, Joseph A. *Dinner in Camelot: The Night America's Greatest Scientists, Writers, and Scholars Partied at the Kennedy White House* (Lebanon, NH: ForeEdge, 2018).

Fernholz, Tim. *Rocket Billionaires: Elon Musk, Jeff Bezos, and the New Space Race* (New York: Houghton Mifflin, 2018).

Glenn, John, and Nick Taylor. *John Glenn: A Memoir* (New York: Bantam Books, 1999).

Guthrie, Julian. *How to Make a Spaceship: A Band of Renegades, an Epic Race, and the Birth of Private Spaceflight* (New York: Penguin Press, 2016).

Kemp, Kenny. *Destination Space: Making Science Fiction a Reality* (London: Virgin Books, 2007).

Koller, Jeffrey W. *The Eden Peace Witness: A Collection of Personal Accounts* (Wichita, KS: Jebeko Publishing, 2004).

Mailer, Norman. *Of a Fire on the Moon* (Boston: Little, Brown, 1970).

McCullough, David. *The Wright Brothers* (New York: Simon & Schuster, 2015).

Meier, Mike, and Mark P. Stucky. *Paragliding: A Pilot's Training Manual* (Orange, CA: Wills Wing, 2004).

Melville, Herman. *Moby-Dick* (New York: Harper & Brothers, 1851).

Paglen, Trevor. *Blank Spots on the Map: The Dark Geography of the Pentagon's Secret World* (New York: Barkley, 2010).

Parker, Alan. *Sid Vicious: No One Is Innocent* (London: Orion, 2008).

Rich, Ben, and Leo Janos. *Skunk Works: A Personal Memoir of My Years at Lockheed* (Boston: Little, Brown, 1994).

Robb, John. *Punk Rock: An Oral History* (London: Ebury Press, 2006).

Rollo, Vera Foster. *Burt Rutan: Reinventing the Airplane* (Baltimore: Maryland Historical Press, 1991).

Sagan, Carl. *Pale Blue Dot* (New York: Random House, 1994).

Salter, James. *The Hunters* (New York: Harper & Brothers, 1956).

Tyson, Neil deGrasse, and Avis Lang. *Accessory to War: The Unspoken Alliance Between Astrophysics and the Military* (New York: W. W. Norton, 2018).

Vance, Ashlee. *Elon Musk: Tesla, SpaceX, and the Quest for a Fantastic Future* (New York: HarperCollins, 2015).

White, Frank. *The Overview Effect: Space Exploration and Human Evolution* (Boston: Houghton Mifflin, 1987).

White, Rowland. *Into the Black: The Extraordinary Untold Story of the First Flight of the Space Shuttle Columbia and the Astronauts Who Flew Her* (New York: Atria Books, 2017).

Wolfe, Tom. *The Right Stuff* (New York: Farrar, Straus and Giroux, 1979).

Yeager, Chuck, and Leo Janos. *Yeager* (New York: Bantam Books, 1985).

ACKNOWLEDGMENTS

I owe a world of gratitude to many people, but first and foremost I want to thank Mark Stucky, who welcomed me into his home and into his life; without his time, his trust, and his patience this book would not have been possible. And to his wife, Cheryl Agin, for being gracious and unwaveringly hospitable.

Michelle Saling recounted the most painful day of her life, and I cannot thank her enough.

High-profile companies in high-risk industries are not often welcoming to journalists. Virgin Galactic nonetheless brought me into the fold. I want to thank Christine Choi for sharing my belief that there was a riveting story to be told, and that I was the right person to tell it; C.J. Sturckow for his initial introduction; Mike Moses for being infinitely candid; Dave Mackay and Todd Ericson for taking me flying. Others shared their stories, their knowledge, and their insights and were instrumental in helping me appreciate the challenges of trying to build a commercial spaceline.

Charlie Bolden, Doug Yurovich, David Barraclough, Doug Jones, and James Cartwright were all generous with their time and helped me better understand my father.

This book grew out of a story that first appeared in the *New Yorker*. Becoming a staff writer at the magazine was the crowning honor of my professional career. I owe David Remnick enormous gratitude for his confidence in me and his patience with this assignment, as it stretched over years. Fabio Bertoni, the general counsel, was ever cool and unflappable; Micah Hauser and Zach Helfand expertly checked the facts. I learn something new from Daniel Zalewski every time we interact; he is the moral conscience of American journalism, full of deep concern for his writers, his readers, his stories, and his subjects. I feel lucky to call him a mentor and a friend.

Many others offered crucial encouragement and support. At an early stage, Martha Raddatz, Nate Fick, and Patrick Radden Keefe all urged me to think about this story as a book; their belief sustained me. Howard Yoon shared sage advice about maintaining narrative tension. Pete Chiarelli offered wisdom about story arcs and thematic coherence. Maria Simon and Neal Katyal provided valuable legal advice. Clara Mulligan shared her premium eye for design. Evan Osnos read the manuscript and gave instrumental feedback. Adam Kushner also read the manuscript with careful consideration; I am grateful for his time and thoughtfulness, and even more so for his friendship.

Writing a book is an oftentimes lonely task. Julia Ioffe, Tom Gjelten, Margaret Angell, Erin Simpson, Victoria Lai, Melissa and Paul Kilbride, Sarabeth Berman, and Patrick Mulligan provided convivial companionship along the way.

A number of institutions also provided support, for which I am immensely grateful. At the Woodrow Wilson International Center for Scholars, Jane Harman and Robert Litwak were welcoming and extremely generous. Kathleen Crown, Joe Stephens, and Margo Bresnen made Princeton University feel like home, one to which I eagerly hope to return someday. What can I say about the Bellagio Center, on the banks of Lake Como? Pilar Palaciá and Alice Luperto host what is arguably the finest writer's retreat on Earth.

I benefited tremendously from the assistance of several researchers, all of whom deserve tremendous thanks. Liz Choi drafted masterful memos. Tim Whitteaker, an archival sleuth, always found the choice detail from an obscure text, and performed the tedious job of transcribing interviews. I relied heavily on Tess Bissell and her editorial instincts; she is brilliant and consistently helped me figure out what I wanted to say. Her suggestions improved each draft of the manuscript immensely.

Zach Helfand, after fact-checking the magazine story, also checked the manuscript. This was not a modest task: he spent months effectively re-reporting the book, with unprecedented rigor and collegiality, and I cannot overstate the significance of his contributions. This is a far better book because of him.

I owe great thanks to Serena Jones, my editor at Henry Holt, for championing me and this project, and for her dedication, support, and partnership throughout the process. And to Madeline Jones, Diana Frost, Chris Sergio, Chris O'Connell, and the rest of the Henry Holt team. My literary agents, Andrew Wylie and Jin Auh, have been boosters, backers, confidants, and collaborators. CAA's Matthew Snyder and Tiffany Ward are always encouraging, and have helped me think about storytelling in novel ways.

This is a book about family, and I feel lucky to be surrounded by such a loving and supportive one. Rebeca has become the fifth member of our family. Ashley is the sister I never had, and Dave has taught me the value of patience. Christian, Jen, Charlie, and Theo fill our lives with laughter and joy. I was blessed to grow up with four incredible grandparents: Ed, Joyce, Mae, and Bob modeled what it meant to be big-hearted and kind, untiring and witty, determined and ambitious, and eternally curious. Joyce, her sister Dot, and Bob continue to inspire the rest of us.

As I explain in the book, my father has had an enormous impact on my life. I am awed by his achievements and his insatiable desire to do more. But it is my mother's radiance, kindness, and empathy that knows no bounds. She is extraordinarily accomplished in her own field but her most lasting impacts are the ones she leaves on every person with whom she interacts. Her consideration, love, and saintly selflessness are qualities I strive to emulate but which prove ever elusive.

My wife and children are everything to me. Watching Oscar and Bohan grow up into wondrous and curious young men, enamored by life's limitless possibilities, is the source of my greatest joy. I adore them, and hope they attain their wildest dreams. This book is for them.

I met Rikki twenty years ago, and have been helplessly in love ever since. She is my most eager reader, dependable editor, receptive therapist, stalwart cheerleader, and best friend. I cannot imagine this book, or my life, without her.

ILLUSTRATION CREDITS

5: Dan Winters
14: Paul Stucky
19: Mark Stucky
31: Mark Stucky
36: Mark Greenberg
45: Chalkie Davies/Getty Images
54: Erik Lassen
68: Don Emmert/AFP via Getty Images
77: Mark Greenberg
85: Michelle LeClerc/Macmillan
90: Jason DiVenere
97: Jason DiVenere
102: Sandy Huffaker/Getty Images
123: Pamela Schmidle
137: NASA
154: Ricky Carioti/The Washington Post/Getty Images
160: National Transportation Safety Board
164: Virgin Galactic
174: Pamela Schmidle
175: Pamela Schmidle
180: James Blair/NASA/JSC
184: Nicholas Schmidle
199: Mike Satren/High Wing Photo Images
208: Dan Winters
212: Nicholas Schmidle
214: Virgin Galactic
231: Cheryl Agin
256: Nicholas Schmidle
259: Virgin Galactic
267: Cheryl Agin
285: Virgin Galactic
286: Virgin Galactic

INDEX

Page numbers in italics indicate figures.